THE SENSORY ORDER

THE SENSORY ORDER

An Inquiry into the Foundations
of Theoretical Psychology

By

F. A. HAYEK

With an Introduction by
HEINRICH KLÜVER

THE UNIVERSITY OF CHICAGO PRESS
CHICAGO · ILLINOIS

The University of Chicago Press, Chicago 60637
The University of Chicago Press, Ltd., London
Copyright in the International Copyright Union

All rights reserved. Originally published 1952
Paperback edition 1976
Printed in the United States of America
04 03 02 01 00 99   6 5 4 3
ISBN: 0-226-32094-4 (paperback)

♾ The paper used in this publication meets the minimum requirements of the American National Standard for Information Sciences—Permanence of Paper for Printed Library Materials, ANSI Z39.48-1992.

PREFACE

A GREAT deal of explanation would be necessary were I to try and justify why an economist ventures to rush in where psychologists fear to tread. But this excursion into psychology has little connexion with whatever competence I may possess in another field. It is the outcome of an idea which suggested itself to me as a very young man when I was still uncertain whether to become an economist or a psychologist. But though my work has led me away from psychology, the basic idea then conceived has continued to occupy me; its outlines have gradually developed, and it has often proved helpful in dealing with the problems of the methods of the social sciences. In the end it was concern with the logical character of social theory which forced me to re-examine systematically my ideas on theoretical psychology.

The paper in which as a student more than thirty years ago I first tried to sketch these ideas, and which lies before me as I write, I was certainly wise not to attempt to publish at the time, even though it contains the whole principle of the theory I am now putting forward. My difficulty then was, as I had been aware even at the time, that though I felt that I had found the answer to an important problem, I could not explain precisely what the problem was. The few years for which I then thought to put the draft away have become a much longer period; and it is little likely that the time will still come when I can devote myself wholly to the working out of these ideas. Yet, rightly or wrongly, I feel that during those years I have learnt at least to state the nature of the problem I had been trying to answer. And as the solution at which I then arrived seems to me to be still new and worth consideration, I have now attempted this fuller exposition of what I had clumsily tried to say in my youthful effort.

The origins of this book, therefore, trace back to an approach to the problem that was current a full generation ago. The psychology which, without much guidance, I read in Vienna in 1919 and 1920, and which led me to my problem, was indeed in all essentials still the psychology of before 1914. Most of the movements which in the interval have determined the direction of psychological research were then either unknown to me or still altogether unheard of: behaviourism (except for the work done in Russia by Pavlov and Bechterev), the gestalt school, or the physiological work of such men as Sherrington or Lashley. And although discussions in Vienna at that time was, of course, full of psychoanalysis, I have to admit that I have never been able to derive much profit from that school. The main authors from which I derived my knowledge were still H. von Helmholtz and W. Wundt, W. James and G. E. Müller, and particularly Ernst Mach. I still vividly remember how in reading Mach, in an experience very similar to that which Mach himself describes with reference to Kant's concept of the *Ding an sich*, I suddenly realized how a consistent development of Mach's analysis of perceptual organization made his own concept of sensory elements superfluous and otiose, an idle construction in conflict with most of his acute psychological analysis.

It was with considerable surprise that, thirty years later, in examining the literature of modern psychology I found that the particular problem with which I had been concerned had remained pretty much in the same state in which it had been when it first occupied me. It seems, if this is not too presumptuous for an outsider to suggest, as if this neglect of one of the basic problems of psychology were the result of the prevalence during this period of an all too exclusively empirical approach and of an excessive contempt for 'speculation'. It seems almost as if 'speculation' (which, be it remembered, is merely another word for thinking) had become so discredited among psychologists that it has to be done by outsiders who have no professional reputation to lose. But the fear of following out complex processes of thought, far from having made discussion more precise, appears to have created a situation in which all sorts of obscure concepts, such as 'representative processes', 'perceptual organization', or 'organized field', are used as if they described definite facts, while actually they stand for somewhat vague theories whose exact content requires to be made clear. Nor has the concentration on those facts which were most readily accessible to observation always meant that attention was directed to what is most important. Neither the earlier exclusive emphasis on peripheral responses, nor the more recent concentration on macroscopic or mass processes accessible to anatomical or electrical analysis, have been entirely beneficial to the understanding of the fundamental problems.

Since this book is concerned with some of the most general problems of psychology, I fear that to many contemporary psychologists it will appear to deal more with philosophical than with psychological problems; but I should be sorry if they should regard it for that reason as falling outside their province. It is true that it presents no new facts; but neither does it employ any hypotheses which are not common property of current psychological discussion. Its aim is to work out certain implications of generally accepted facts or assumptions in order to use them as an explanation of the central problem of the nature of mental phenomena. Indeed, if the views generally held on the subject are approximately true, it would seem as if something of the kind here described must happen, and the surprising fact would seem that so little attempt has been made to work out systematically these consequences of existing knowledge. Perhaps such an effort of effectively thinking through these implications requires a combination of qualifications which nobody possesses to a sufficient degree and which the specialist who feels sure in his own field therefore hesitates to undertake. To do it adequately one would indeed have to be equally competent as a psychologist and as a physiologist, as a logician and as a mathematician, and as a physicist and as a philosopher. I need scarcely say that I possess none of these qualifications. But since it is doubtful whether anybody does, and since at least nobody who possesses them as yet has tried his hand at this problem, it is perhaps inevitable that the first attempt should be made by somebody who had to try and acquire the necessary equipment as he went along. A satisfactory execution of the thesis which I have outlined would probably require the collaboration of several specialists in the different fields.

The parts of the problem on which I feel tolerably confident that I have something of importance to say are the statement of the problem, the general principles of its solution, and some of the consequences which follow from the latter for epistemology and the methodology of the sciences. The sections of the book with which I am therefore tolerably satisfied are the beginning and the end: Chapters I and II and Chapters VII and VIII. Perhaps it would have been wiser if I had made no attempt to implement the programme outlined in the earlier chapters, since the central part of the book in which this is attempted is unavoidably both more technical and more amateurish than the rest. Yet it seemed important to illustrate the general principles stated in the earlier chapters by some attempt at elaboration, even at the risk of slipping at particular points. In some ways this would not greatly matter: I am much more concerned about what would constitute an explanation of mental phenomena, than whether the details of this theory are entirely correct. Since we are still in a position where we are not certain what would constitute an explanation, any theory which, if it were correct, would provide one would be a gain, even if it should not be tenable in all respects.

Even the present version of this book has occupied me for several years, and though I have endeavoured to acquaint myself with the relevant literature, I am not sure that I have been able fully to keep up with current developments. It seems as if the problems discussed here were coming back into favour and some recent contributions have come to my knowledge too late to make full use of them. This applies particularly to Professor D. O. Hebb's *Organization of Behaviour* which appeared when the final version of the present book was practically finished. That work contains a theory of sensation which in many respects is similar to the one expounded here; and in view of the much greater technical competence of its author I doubted for a while whether publication of the present book was still justified. In the end I decided that the very fullness with which Professor Hebb has worked out the physiological detail has prevented him from bringing out as clearly as might be wished the general principles of the theory; and as I am concerned more with the general significance of a theory of that kind than with its detail, the two books, I hope, are complementary rather than covering the same ground.

I owe a debt of deep gratitude to the London School of Economics and the Committee on Social Thought of the University of Chicago for giving me the leisure to devote so much time to problems which lie outside the field where my main duties lie. To my friends Karl R. Popper and L. von Bertalanffy and to Professor J. C. Eccles I am much indebted for reading and commenting upon earlier drafts of this book. And without the acute criticism of the manuscript by my wife the book would contain even more obscurities and so slovenly expressions than it undoubtedly still does.

F. A. HAYEK

CONTENTS

CONTENTS

CONTENTS

INTRODUCTION

by HEINRICH KLÜVER

I⊤ has been said that a philosopher is a man who has a bad conscience whenever he hears the word philosophy. Nowadays psychologists no longer seem to develop feelings of guilt when encountering the word psychology. This state of affairs can certainly not be accounted for by assuming that the whole field of psychology has suddenly acquired the status of a 'science'. In fact, scientific progress in psychology within the last generation according to some critics has been deplorable. A turn for the better, however, would not necessarily be achieved by eliminating all psychologists. There is no doubt that physiologists, neurologists, psychiatrists, anatomists, sociologists, biologists, and workers in other fields would keep psychology alive in one form or another if psychologists were to disappear from the contemporary scene. Investigators in various non-psychological fields are, in the pursuit of their inquiries, again and again forced to deal with psychological problems and even driven to consider problems of theoretical psychology. For example, about ten years or so ago, Sherrington felt compelled to consider the interrelations of neurophysiological and psychological phenomena and to devote several hundred pages in his *Man on His Nature* to an examination of problems of the 'mind'.

Dr Hayek's book, which represents an analysis of the sensory order in relation to problems of theoretical psychology, raises the question whether the time has not again come for psychologists to develop, at least occasionally, a bad conscience when hearing the word psychology. On the one hand, there seems to have been a decline in the quality and quantity of systematic endeavours in the field of theoretical psychology during the last decades; on the other hand, the multifarious activities of psychologists seem to make it more than ever necessary to find a common point of reference. As long as it is assumed, or as long as the illusion is to be kept alive, that the diverse activities of psychologists involve a common factor referrred to by the word psychology, the general conceptual framework of such a psychology must remain of fundamental interest. There is, of course, no lack of theorizing in modern psychology. It is one thing, however, to develop a theory based on the detailed experimental analysis of a particular problem; it is another thing to examine the conceptual tools of theoretical psychology itself. Even particular theories do not always escape the complexity of matters psychological. When G. E. Müller summarized his fifty years of efforts in the field of colour vision, it took him about 650 pages to present his theory of colour vision and he insisted that any simpler theory could be formulated only at the price of ignoring relevant facts. When it comes to systematic efforts in the field of theoretical psychology, it has become increasingly obvious in recent years that psychologists find their particular tasks (ranging from an analysis of ocular to an analysis of political movements and from investigations of the sexual behaviour of male army ants to that of human females) so all-absorbing, time-consuming and exacting that they rarely seem to do anything but increase the number of hastily conceived and irresponsible theories. In fact, nowadays, only a man like Dr Hayek who is sufficiently removed from the noisy market places of present-day psychology appears to have the necessary detachment and peace of mind for a systematic inquiry into the foundations of theoretical psychology.

It is fortunate indeed that Dr Hayek has chosen the sensory order as a basis for discussing problems of theoretical psychology. More than a century ago, in 1824, Flourens insisted that *une anatomie sans physiologie serait une anatomie sans but*. There is no doubt that an 'anatomy without physiology' as well as a 'physiology without anatomy' are still with us. Even at the present time, it is not difficult to find books on the 'physiology' of the nervous system which are in effect nothing but books on 'anatomy' containing elaborate physiological footnotes. The relations between physiology and anatomy acquire particular significance and complexity when it comes to the field of sensory physiology. If not the last, at least the last monumental attempt to cope with problems of the sensory order is to be found in J. von Kries's *General Sensory*

Physiology published in 1923. It was von Kries who explicitly stated what has been recognized by practically all investigators in this field, namely, that sensory physiology is different from all other fields of physiology and, in fact, from all other natural sciences in that its problems are intimately related to, if not identical with, certain problems of psychology, epistemology, and logic. In fact, sensory physiology and sensory psychology are to a great extent indistinguishable. The psychologist will remember that the duplicity theory of vision formulated by von Kries has stood the test of time longer than is generally the case with scientific theories and that Selig Hecht, only a few years ago, when performing energy measurements to determine the minimum energy necessary for vision, found values of the same order of magnitude as were found by von Kries more than forty years ago although von Kries did not even make energy measurements. Since Hecht considered it 'astonishing to see the admirable way' in which von Kries accomplished this task, he felt called upon to pay tribute to von Kries's skill and care in the evaluation of absorptions, reflections, lens factors, and the like, which are necessary in determining the minimum energy. It cannot be said, therefore, that the man who insisted that problems of sensory physiology cannot be adequately treated without recourse to psychology, epistemology, and logic did not have the necessary 'hard-boiled' attitude in scientific matters; the converse is obviously true. Just why the sustained and disciplined thinking of a 'hard-boiled' professor of physiology in matters of sensory physiology should be dismissed as 'mere' philosophy by psychologists is a problem which clearly requires an analysis by a competent historian unless one assumes that the inability of most psychologists to handle logical and epistemological tools explains such a phenomenon. Such phenomena, unfortunately, have not been rare in the history of psychology. Ziehen, the neuroanatomist, psychiatrist, psychologist, and logician, who wrote a textbook of physiological psychology that went through numerous editions, also wrote an 'epistemology on psychophysiological and physical basis' in 1913. It is true that this book has a forbidding title and about 600 pages; but it is probably also true that no psychologist alive ever read all of its pages in the period between World War I and World War II. There is no doubt that a critical examination of the concepts of general sensory physiology in relation to psychology and other fields is a prerequisite for further progress in many physiological and psychological areas of investigation. Dr Hayek, who appears to be far too modest in evaluating his own competence in handling and elucidating sensory physiological and psychological concepts, is, therefore, performing a task urgently needed for further scientific progress.

Problems of the sensory order and the relations between physical and sensory phenomena have been of perennial interest not only to psychologists and physiologists, but also to mathematicians, logicians, and physicists. Recently they have even become of interest to communication engineers. As P. du Bois-Reymond once pointed out, all of us are enclosed 'in the box of our perceptions'. There have always been some who think that it is possible to escape from this box and there have always been others who think that this is not possible. Ziehen, for instance, was of the opinion that we *find* everywhere identities, similarities, and differences in examining the 'given', i.e., the raw data furnished by experience or, to use his expression, the 'gignomene'. A fundamental classificatory principle is thus part and parcel of the 'gignomene' themselves and constitutes an ultimate 'unexplainable and indefinable fact'. It is of interest that von Kries, too, believed that the existence of similarities is an ultimate fact neither requiring nor permitting of an explanation. The view, however, that the real content of experience resisting any further analysis is to be found in sensory phenomena has always clashed with the view that the persistence of fixed functional relations between these phenomena constitutes the content of true reality.

In a brief space, it is impossible to outline even the essentials of Dr Hayek's theory, but from a broad point of view his theory may be said to substantiate Goethe's famous maxim 'all that is factual is already theory' for the field of sensory and other psychological phenomena. According to Dr Hayek, sensory perception must be regarded as an act of classification. What we perceive are never unique properties of individual objects, but always only properties which the objects have in common with other objects. Perception is thus always an interpretation, the placing of something into one or several classes of objects. The characteristic attributes of sensory qualities, or the classes into which different events are placed in the process of perception, are not attributes which are possessed by these events and which are in some manner 'communicated' to the mind; they consist entirely in the 'differentiating' responses of the organism by which the qualitative classification or order of these events is created; and it is contended that this classification is based on the connexions created in the nervous system by past 'linkages.' The qualities which we attribute to the experienced objects are, strictly speaking, not properties of objects at all, but a set of relations by which our nervous system classifies them. To put it differently, all we know about the world is of the nature of theories and all 'experience' can do is to change these theories. All sensory perception is necessarily 'abstract' in that it always selects certain aspects or features of a given situation. Every sensation, even the 'purest,' must therefore be regarded as an interpretation of an event in the light of the past experience of the individual or the species. Experience operates on physiological events and arranges them into a structure or order which becomes the basis of their 'mental' significance. In the course of ontogenetic or phylogenetic development, a system of connexions is formed which records the relative frequency with which different groups of internal and external stimuli have acted together on the organism. Each individual impulse or group of impulses on its occurrence evokes other impulses which correspond to stimuli which in the past have usually accompanied its occurrence. The primary impulse through its acquired connexions will set up a bundle of secondary impulses, a 'following' of the primary impulse. It is the total or partial identity of this 'following' which determines different forms of classification. The essential characteristic of the order of sensory qualities is that, within that order, each stimulus or group of stimuli does not possess a unique significance represented by particular responses, but that the stimuli are given different significance if they occur in combination with, or are evaluated in the light of, an infinite number of other stimuli which may originate from the external world or from the organism itself. A wide range of mental phenomena, such as discrimination, equivalence of stimuli, generalization, transfer, abstraction, and conceptual thought, may all be interpreted as different forms of the same process of classification which is operative in creating the sensory order. The fact that this classification is determined by the position (in a topological, not a spatial, sense) of the individual impulse or group of impulses in a complex structure of connexions, extending through a hierarchy

of levels, has important consequences when it comes to considering the effects of physiological or anatomical changes.

These formulations of the author must suffice to characterize at
least some aspects of the theory presented in his book. Investigators concerned with an analysis of the logical structure of
natural sciences have insisted that the transition from concepts of
'substance' to concepts of 'function' is characteristic of the historical development of science. 'Thing-concepts' have gradually and
often painfully yielded to 'relational concepts'. Even Freud, some
critics have insisted, is still a 'substance' thinker. In this connexion Dr Hayek's theory appears very modern indeed since not
even traces of 'thing-concepts' are left in his theory. 'Mind' for
him has turned into a complex of relations; it is simply 'a particular order of a set of events taking place in some organism and in
some manner related to, but not identical with, the physical order
of events in the environment'. In addition, his theory, perhaps
more than any other, emphasizes the far-reaching importance of
'experience' and 'learning'. Certain theories have always stressed
the factor of 'experience' while others have stressed the importance of the conditions, structures, or presuppositions which
make experience possible. The *relations* between these two sets of
factors, however, present peculiar difficulties. In elucidating the
complexity of these relations, Dr Hayek probably makes his most
important and original contributions. It has been said that there
are no permanent or fixed 'objects', but only ways of knowing
'objectively'. The implication of the theory presented here is that
even the ways of knowing 'objectively' are not stable, or only
relatively stable, and that the ordering principles themselves are
subject to change. Dr Hayek, therefore, does not take a static
view of either the 'elements' or the 'relational' structure involved
in the sensory or any other kind of order. Conceptual thinking, as
he rightly emphasizes, has long been recognized as a process of
continuous reorganization of the (supposedly constant) elements
of the phenomenal world. In his opinion, however, there is no
justification for the sharp distinction between the more abstract
processes of thought and direct sensory perception since the qualitative elements, of which the phenomenal world is built up, and
the whole order of the sensory qualities are themselves subject to
continuous change. The fact that there can be nothing in our
mind which is not the result of ontogenetically or phylogenetically

established 'linkages' is not meant to exclude processes of re-classification. At the same time it is to be clearly understood that at least a certain part of what we know at any moment about the external world is not learned by sensory experience, but is rather implicit in the means through which we can obtain such experience; that is, it is determined by the order of the previously established apparatus of classification. To express it differently, there is, on every level, a part of our knowledge which, although it is the result of experience, cannot be controlled by experience because it constitutes the ordering principle. In considering the implications of Dr Hayek's theory, the impression is gained that not only the characteristics and properties of the organism involved in 'classifying' activities but also the characteristics of the 'environment' appear in a new light. Man occupies only a small corner of the terrestrial biosphere, including the recently developed, chemically highly active, and man-made anthroposphere of A. P. Pavlov. If pre-sensory and sensory 'linkages' are formed not only during the life of the individual, but also in the course of phylogenetic development, the characteristics of the environment, in which the building-up of the apparatus of classification or orientation occurs, assume special importance. If the apparatus of classification is shaped by the conditions in the environment in which we live and if it represents a kind of map or reproduction of relations between elements of this environment, the question arises as to the extent to which environmental factors 'colour' or 'condition' principles of ordering. Perhaps Vernadsky's bio-geochemistry has, in the light of Dr Hayek's theory, unexpected psychological implications. In the meantime, the striking results on 'conditioned sensations' recently obtained by Ivo Kohler have demonstrated how strongly environmental factors and conditions may influence sensory phenomena during the life of an individual.

It is not possible to comment in detail on the skill and knowledge with which Dr Hayek has utilized psychological, physiological, and other data to support his thesis and to enumerate the many problems and theories upon which his penetrating analysis has significant bearing. His concepts of the 'map', the 'model', and related concepts appear to be promising tools in analysing brain mechanisms and behaviour. What is perhaps most pertinent is that his theory suggests definite lines of experimentation. For instance, it should be possible not only to change sensory qualities experimentally, but to create altogether new sensory qualities which have never been experienced before. The psychologist is likely to find this theory helpful in devizing new experiments even beyond the scope indicated by the author himself. In considering the consequences and implications of his own theory and in trying to define its content as unambiguously as possible, the author does not hesitate to point out that an experimental confirmation of theories, such as Semon's 'engram' theory or Paul Weiss's 'resonance' theory, would disprove his own theory.

A great historian once expressed the view that 'no man, and no product of all a man's labour either, is like a perfectly thought-out book, and merely to point out lacunae in some pages and deficiences in others must seem much more an envious job of rival contemporaries than a historian's true duty . . .' When viewing the complex structure of a psychological theory, envious 'rival contemporaries' undoubtedly will always try, no matter how difficult the job may be, to establish that certain *petits faits signicatifs* or even merely 'little facts' have not been built into the edifice. About twenty years ago, Lashley, in commenting on an experimental investigation concerned with analysing the interdependence of relations and relata and establishing forms of invariance in behaviour, spoke of 'the tracing of relations through the intricate web of dependent processes which is "mind"'. Dr Hayek has done more than his share in tracing relations through the intricate web of 'mind'. His is one of the most interesting and significant books on theoretical psychology that has appeared in this country during the last decades.

CHAPTER I

THE NATURE OF THE PROBLEM

1. WHAT IS MIND?

1.1. The nature of the subject of this study makes its first task the most important and the most difficult: clearly to state the problem to which it will attempt an answer. We shall have moved a considerable distance towards the solution of our problem when we have made its meaning precise and have shown what kind of statement could be regarded as a solution.

1.2. The traditional heading under which our problem has been discussed in the past is that of the 'relation' between mind and body, or between mental and physical events. It can also be described by the questions of 'What is mind?' or 'What is the place of mind in the realm of nature?' But while these expressions indicate a general field of inquiry, they do not really make it clear what it is that we want to know. Before we can successfully ask how two kinds of events are related to each other (or connected with each other), we must have a clear conception of the distinct attributes by which they can be distinguished. The difficulty of any fruitful discussion of the mind-body problem consists largely in deciding what part of our knowledge can properly be described as knowledge of mental events as distinguished from our knowledge of physical events.

1.3. We shall attempt to avoid at first at least some of the difficulties of this general problem by concentrating on a more definite and specific question. We shall inquire how the physiological impulses proceeding in the different parts of the central nervous system can become in such a manner differentiated from each other in their functional significance that their effects will differ from each other in the same way in which we know the effects of the different sensory qualities to differ from each other. We shall have established a 'correspondence' between particular physiological events and particular mental events if we succeed in showing that there can exist a system of relations between these physiological events and other physiological events which is identical with the system of relations existing between the corresponding mental events and other mental events.

1.4. We select here for examination the problem of the determination of the order of sensory qualities because it seems to raise in the clearest form the peculiar problem posed by all kinds of mental events. It will be contended that an answer to the question of what determines the order of sensory qualities constitutes an answer to all questions which can be meaningfully asked about the 'nature' or 'origin' of these qualities; and further, that the same general principle which can be used to account for the differentiation of the different sensory qualities serves also as an explanation of the peculiar attributes of such other mental events as images, emotions, and abstract concepts.

1.5. For the purposes of this discussion we shall employ the term sensory 'qualities' to refer to all the different attributes or dimensions with regard to which we differentiate in our responses to different stimuli. We shall thus use this term in a wide sense in which it includes not only quality in the sense in which it is contrasted with intensity, extensity, clearness, etc., but in a sense in which it includes all these other attributes of a sensation.[1] We shall speak of sensory qualities and the sensory order to distinguish these from the affective qualities and the other mental 'values' which make up the more comprehensive order of 'mental qualities'.

2. THE PHENOMENAL WORLD AND THE PHYSICAL WORLD

1.6. A precise statement of the problem raised by the existence of sensory qualities must start from the fact that the progress of the physical sciences has all but eliminated these qualities from our scientific picture of the external world.[2] In order to be able to

[1] See E. G. Boring 1933, pp. 22–23 and 1942, p. 42.
[2] Cf., e.g., M. Planck, 1926, p. 5 : 'The sense perceptions have been definitely eliminated from physical acoustics, optics and heat. The physical definitions of sound, colour, and temperature are to-day in no way associated with the immediate perception of the respective senses, but sound and colour are defined respectively by the frequency and wavelength of oscillations, and temperature is measured theoretically on the absolute temperature scale corresponding to

give a satisfactory account of the regularities existing in the physical world the physical sciences have been forced to define the objects of which this world exists increasingly in terms of the observed relations between these objects, and at the same time more and more to disregard the way in which these objects appear to us.

1.7. There exist now, in fact, at least two[1] different orders in which we arrange or classify the objects of the world around us: one is the order of our sense experiences in which events are classified according to their sensory properties such as colours, sounds, odours, feeling of touch, etc.; the other is an order which includes both these same and other events but which treats them as similar or different according as, in conjunction with other events, they produce similar or different other external events.

1.8. Although the older branches of physics, particularly optics and acoustics, started from the study of sensory qualities, they are now no longer directly concerned with the perceptible properties of the events with which they are dealing. Nothing is more characteristic of this than the fact that we find it now necessary to speak of 'visible light' and 'audible sound' when we want to refer to the objects of sense perception. To the physicist 'light' and 'sound' now are defined in terms of wave motions, and in addition to those physical events, which, as is true of certain ranges of 'light' and 'sound' waves, cause definite sense experiences, he deals with imperceptible events like electricity, magnetism, etc., which do not directly produce specific sensory qualities.[2]

1.9. Between the elements of these two orders there exists no simple one-to-one correspondence in the sense that several objects or events which in the one order belong to the same kind or class will also belong to the same kind or class in the other order. They

the second law of thermodynamics.' See also 1949 (1941), p. 108. On the fact that this applies not only to the 'secondary' qualities see H. Margenau, 1950, pp. 7 and 49.

[1]Since, as we shall see, the movement from the sensory to the physical order is a gradual one, there is, strictly speaking, an infinite range of such orders of which the naïve picture of the sensory world and the latest scientific knowledge are merely the most significant types.

[2]Cf., J. von Kries, 1923, p. 67, and E. G. Boring, 1942, p. 97. As late as 1935 the latter author could still write (p. 236) that 'it is the traditional view of psychology that the attributes of sensation show a one-to-one correspondence to the dimensions of the stimulus.'

constitute different orders precisely because events which to our senses may appear to be of the same kind may have to be treated as different in the physical order, while events which physically may be of the same or at least a similar kind may appear as altogether different to our senses.

1.10.   These two orders have been variously described by different authors as the subjective, sensory, sensible, perceptual, familiar, behavioural or phenomenal[1] world on the one hand, and and as the objective, scientific, 'geographical', physical, or sometimes 'constructional' on the other. In what follows we shall regularly employ the pair of terms 'phenomenal' and 'physical'[2] to describe the order of events perceived in terms of sensory qualities and the order of events defined exclusively in terms of their relations respectively, although we shall occasionally employ the term 'sensory' as equivalent to phenomenal, especially (as in the title of this book) in the phrase 'sensory order'. We shall later (Chapters V and VIII) also describe these two orders as the 'macrocosm' and the 'microcosm' respectively. Their relation is the central problem of this book.

1.11.   It is important not to identify the distinction between the phenomenal and the physical order with the distinction between either of these and what in ordinary language is described as the 'real' world. The contrast with which we are concerned is not between 'appearance' and 'reality' but between the differences of events in their effects upon each other and the differences in their effects on us. It is indeed doubtful whether on the plane on which we must examine these problems the term 'real' still has any clear meaning.[3] For the purposes of our discussion, at any rate, we shall not be interested in what a thing 'is' or 'really is' (whatever that may mean), but solely in how a particular object or event differs from other objects or events belonging to the same order or universe of discourse. It seems that a question like 'what is x?' has meaning only within a given order, and that within this limit it must always refer to the relation of one particular event to other

[1]In German often by the not fully translatable word *anschaulich*.
[2]To prevent confusion it should perhaps be pointed out that the 'physical language' of the logical positivists refers to the phenomenal and not to the physical order.
[3]These doubts have not been dispelled by the careful distinction of various kinds and degrees of 'reality' (*Wirklichkeit*) by W. Metzger, 1941, Chapter 2.

4

events belonging to the same order. We shall see that the mental and the physical world are in this sense two different orders in which the same elements can be arranged; though ultimately we shall recognize the mental order as part of the physical order, a part, however, whose *precise* position in that larger order we shall never be able to determine.

1.12. Historically the concept of the 'real' has been formed in contradistinction to mere 'illusions' based on sense deceptions or on other experiences of purely mental origin. There is, however, no fundamental difference between such corrections of one sense experience by others, as we employ, e.g., to discover an optical illusion, and the procedure employed by the physical sciences when they ascertain that two objects which may to all our senses appear to be alike do not behave in the same way in relation to others. To accept this latter test as the criterion of 'reality' would force us to regard the various constructs of physics as more 'real' than the things we can touch and see, or even to reserve the term 'reality' to something which by definition we can never fully know. Such a use of the term 'real' would clearly pervert its original meaning and the conclusion to be drawn from this is probably that it should be altogether avoided in scientific discussion.[1]

1.13. The relation between the physical and the phenomenal order raises two distinct but related problems. The first of these problems presents the task of the physical sciences while the second creates the central problem of theoretical psychology. The task of the physical sciences is to replace that classification of events which our senses perform but which proves inadequate to describe the regularities in these events, by a classification which will put us in a better position to do so. The task of theoretical psychology is the converse one of explaining why these events, which on the basis of their relations to each other can be arranged in a certain (physical) order, manifest a different order in their effect on our senses.

1.14. The problems of the physical sciences arise thus from the fact that objects which appear alike to us do not always prove to behave in the same way towards other objects; or that objects which phenomenally resemble each other need not be physically

[1]On the gradual evolution of the scientific world picture from the efforts of the child, and on the use of the term 'real', see M. Planck, 1949 (1941), especially pp. 90 and 95–105.

similar to each other, and that sometimes objects which appear to us to be altogether different may prove to be physically very similar.

1.15. It is this fact which has made it necessary, in order to build up a science capable of predicting events, to replace the classification of objects or events which our senses effect by a new classification which corresponds more perfectly to the manner in which those objects or events resemble or differ from each other in the effects which they have upon each other. But this progressive substitution of a purely relational for a qualitative or sensory order of events provides the answer to only one part of the problem which is raised by the existence of the two orders. Even if we had fully answered this problem we should still not know why the different physical objects appear to us as they do.

1.16. It is because the physical sciences have shown that the objects of the external world do not regularly differ in their effects upon each other in the same way in which they differ in their effects upon our senses that the question why they appear to us as they do becomes a legitimate problem and indeed the central problem of theoretical psychology. In so far as the similarities or differences of the phenomena as perceived by us do not correspond with the similarities or differences which the perceived events manifest in their relations to each other, we are not entitled to assume that the world appears to us as it does because it is like that; the question why it appears to us as it does becomes a genuine problem.[1]

1.17. It is, perhaps, still true that psychologists in general have not yet become fully aware of the fact that, as a result of the development of the physical sciences, the explanation of the qualitative order of the phenomenal world has become the exclusive task of psychology. What psychology has to explain is not something known solely through that special technique known as 'introspection', but something which we experience whenever we learn anything about the external world and through which indeed we know about the external world; and which yet has no place in our scientific picture of the external world and is in no way explained by the sciences dealing with the external world: qualities. Whenever we study qualitative differences between experiences we are studying mental and not physical events, and much that we be-

[1] Cf., K. Koffka, 1934, pp. 75ff.

lieve to know about the external world is, in fact, knowledge about ourselves.[1]

1.18. It is thus the existence of an order of sensory qualities and not a reproduction of qualities existing outside the perceiving mind which is the basic problem raised by all mental events. Psychology must concern itself, in other words, with those aspects of what we naïvely regard as the external world which find no place in the account of that world which the physical sciences give us.

1.19. This reformulation of the central problem of psychology has thus been made necessary by the fact that the physical sciences, even in their ideal perfect development, give us only a partial explanation of the world as we know it through our senses and must always leave an unexplained residue. After we have learnt to distinguish events in the external world according to the different effects they have upon each other, and irrespective of whether they appear to us as alike or different, the question of what makes them appear alike or different to us still remains to be solved. The empirical establishment of correspondences between certain phenomenal and certain physical constellations of events is no sufficient answer to this question. We want to know the kind of process by which a given physical situation is transformed into a certain phenomenal picture.

1.20. Since the peculiar order of events which we have called the phenomenal order manifests itself only in the responses of certain kinds of organisms to these events, and not in the relation of those events to each other, it is natural to search for an explanation of this order in some feature of the structure of these organisms. We shall eventually find it in the fact that these organisms are able within themselves to reproduce (or 'build models of') some of the relations which exist between the events in their environment.

1.21. The fact that the problem of psychology is the converse of the problem of the physical sciences means that while for the latter the facts of the phenomenal world are the data and the order of the physical world the *quaesitum*, psychology must take the physical world as represented by modern physics as given and try to reconstruct the process by which the organism classifies the physical events in the manner which is familiar to us as the order of sensory qualities. In other words: psychology must start from

[1]Cf., F. A. Hayek, 1942, p. 279.

stimuli defined in physical terms and proceed to show why and how the senses classify similar physical stimuli sometimes as alike and sometimes as different, and why different physical stimuli will sometimes appear as similar and sometimes as different.[1]

3. STIMULUS, IMPULSE, AND THE THEORY OF THE SPECIFIC ENERGY OF NERVES

1.22.   Before we proceed farther it is necessary to define more precisely some of the terms we shall have constantly to employ. This applies especially to the terms 'stimulus' and 'nervous impulse' and more particularly to the sense in which we shall speak of particular 'kinds' of stimuli or of the same and of different nervous impulses. It will be convenient also to consider already at this stage the meaning and significance of the famous principle of the 'specific energy of nerves.'

1.23.   The term *stimulus* will be used throughout this discussion to describe an event external to the nervous system which causes (through or without the mediation of special receptor organs) processes in some nerve fibres which by these fibres are conducted from the point at which the stimulus acts to some other point of the nervous system. It appears that at least some receptor organs are sensitive not to the continuous action of any one given stimulus but only to changes in that stimulus. Whatever it is that is produced in the nerve fibre and propagated through it we shall call the *impulse*.

1.24.   The physical event acting as a stimulus is described as such only with regard to its action on the receptors.[2] This leads sometimes to a rather confusing distinction between the stimulus and its 'source', sometimes described as the stimulus object. What will here be described as stimulus will always be the proximal stimulus,[3] i.e., the last known physical event in the chain which leads to the production of the impulse. In some instances (particularly in the case of odours) this proximal physical stimulus, however, is

[1] Cf., E. G. Boring, 1942, p. 120 : 'Nowadays we consider first the dimensions of the stimulus, and then seek to discover what phenomenal consequences they yield. We used to inquire about the physical causes of hue: now we ask about the effects of monochromatic light.'
[2] R. S. Woodworth, 1938, p. 451.
[3] K. Koffka, 1935, p. 80.

not certainly known, and we must be satisfied with reference to some more remote event which has then to be regarded as the source of an unknown proximal stimulus.

1.25. It is necessary from the outset carefully to avoid the assumption that to each kind of sensation there will always correspond *one* stimulus of a particular kind. Not only can several different stimuli produce the same sensation, but it appears that in many instances, and perhaps as a rule, several different stimuli, acting on different receptors, may be required to produce a particular sensation.[1]

1.26. Since our central problem is the manner in which different stimuli affect our nervous system, or how they are classified by it, we clearly cannot make our starting point that classification of the stimuli which our senses perform. The distinction between different stimuli, or between different kinds of stimuli, must be independent of the different effects they have on the organism. This independence can never be complete, since all our knowledge of the external events is derived from our sensory experience. But it can be independent in the sense that we can classify the stimuli not according to their direct effects on our senses, but according to the effects which they exercise on other external events, which in turn act as stimuli on our senses. This classification of the events which act as stimuli, according to their effects on other events which in turn are classified according to their effects on still others, is, of course the classification of the stimuli developed by the physical sciences; and it is this which we must adopt.

1.27. We shall, e.g., have to regard as the same physical stimulus not all light which appears to us to have the same colour, or all substances which smell alike, but only light waves which in various combinations with other physical objects (usually apparatus designed for the purpose) produce the same effects, or substances which in their chemical composition are identical.

1.28. For our purpose it will also be necessary to regard as different any stimuli which are physically identical but which act on different parts of the body, since it is by no means obvious (or always true) that such stimuli should produce the same sensory qualities. The question why as a rule stimulation of different individual receptors by physically identical stimuli should produce

[1] C. T. Morgan, 1943, pp. 297–8.

9

similar sensations is in fact the simplest form in which our problem arises.

1.29. The production of a nervous impulse by a stimulus is usually mediated by the selective action of specific receptor organs which respond to certain kinds of stimuli but not to others. This selectivity of the receptor organs is, however, not perfect. Even the so-called 'adequate stimuli' to which a given receptor normally responds, consist as a rule not only of one precisely defined physical stimulus (such as, e.g., waves of a particular frequency) but to a more or less wide range of such stimuli extending, e.g., over a certain band of frequencies. In addition to this, some events other than the adequate stimuli can often set up impulses in a given nerve fibre. An impulse in the visual nerves and the consequent sensation of light may, for instance, be caused by a blow on the eyeball.

1.30. The receptor organs thus already perform a certain sorting out, or classification, of the stimuli, and there will be no strict correspondence between the different stimuli and the different impulses. Moreover, only a small part of the physical events in our environment are capable of acting as stimuli or are recorded by impulses in the nerve fibres. Of the continuous range of electromagnetic waves only a very small band acts on our organs of vision while by far the greater part of this range does not act as a stimulus on our nerves.

1.31. Impulses in a particular sensory nerve fibre may thus be set up by any one of a group of stimuli which physically may be similar or altogether different. But if a given fibre responds to any of these stimuli, the character of the impulse transmitted will always be the same, irrespective of the nature of the stimulus. The effect of the impulse is independent of the nature of the particular kind of stimulus which evokes it, and any characteristic effects which this particular impulse brings about must therefore be due to something connected with that impulse and not to any attributes of the stimulus.

1.32. This is the main contention of the so-called principle of the specific energy of nerves. When it was first enounced by Johannes Müller, it was aimed against the conception that the nervous impulses transmitted some attribute of the stimulus to the brain; and it was intended to emphasize that the sensation produced depended solely on the fibre which carried the impulse and not on what had caused that impulse. The form in which it was stated, however, was not free from ambiguity and soon gave rise to a new misconception.

1.33. The fact that the theory was called the theory of the specific energy of nerves led to its being connected with one particular alternative explanation of the determination of sensory qualities which is no less questionable than the theory which it was intended to displace. On this interpretation it was understood to mean that, if it was not the physical properties of the stimuli which determined the quality of the resulting sensations, it must be some property of the individual impulses proceeding in the different fibres which in some sense 'corresponds' to the differences of the sensory qualities.

1.34. Although this is by no means a necessary consequence of the proposition which Johannes Müller had been anxious to establish, it was widely assumed, in fact, that the sensory qualities produced by impulses in different fibres would be different, similar or equal according as the physical properties of the corresponding impulses differed from or resembled each other. This interpretation was to some extent suggested by Müller's own formulation of the theory in which he asserted more than was necessary to establish his conclusions. In his summary of his theory he stated that 'the sensation is not the conduction of a quality or state of an external body to the consciousness, but the conduction to the consciousness *of a quality or state of our sensory nerves* induced by an external cause';[1] and he went on to emphasize that these qualities are different with the different senses.

1.35. The recognition, however, that the difference of the sensory qualities is not due to the communication of a difference in the stimuli does by no means make the conclusion inevitable that it must then be a difference in the properties of the impulses taking place in the different fibres, which accounts for them. To interpret the theory of the specific energy of nerves in this sense is merely to accept at this stage an explanation similar to that rejected at the

---

[1]Johannes Müller, 1838, I. p. 780 and II, p. 262. What we regard as the illegitimate interpretation of the theory of the specific energy of nerves was later explicitly formulated by G. E. Müller (1896) in the second of his five 'psychophysical axioms' (see E. G. Boring, 1942, p. 89) and became widely known mainly in the form in which it was expounded by H. Hering, 1885, 1913. The basic idea has recently been revived by P. Weiss, 1941, and R. W. Sperry, 1945.

earlier stage: the specific character of the effect of a particular impulse need be neither due to the attributes of the stimulus which caused it, nor to the attributes of the impulse, but may be determined by the position in the structure of the nervous system of the fibre which carries the impulse.[1]

1.36. We do not only possess no information which would entitle us to assume that the impulses carried by the different fibres differ qualitatively, but, what is more important, it also seems impossible to conceive of such differences between the physical attributes of the individual impulses that they could be said in any sense to 'correspond' to the differences of the sensory qualities. Even if qualitative differences between the impulses were discovered, this would not yet provide an answer to our problem. It would still be necessary to show how these differences in quality determined the different effects which the different impulses exercise upon each other; and while it is conceivable that these latter differences may be connected with differences in their individual physical attributes, this need not be so. The important point here is that no differences of the individual impulses as such would provide an explanation of the differences between their mental equivalents, and that any differences of their causal connexions with each other seem at least as likely to be due to structural connexions as to qualitative affinities. This is important especially because the hope of thereby providing an explanation of the differences in mental qualities appears to have been the prime motive for the persistent and unsuccessful search for 'specific energies', and because the same conception seems also largely responsible for the persistence of the belief in a 'pure core' of sensation.[2]

1.37. The evidence which we possess suggests, in fact, that the impulses carried by the different fibres, at least within any one sense modality, are qualitatively identical, so that, if we were to cut two sensory fibres and to re-connect the lower part of each with the upper part of the other, they would still function but exchange the results which an impulse in either would cause. It seems, therefore, that the cause of the specific effects of the impulses in different fibres must be sought, not in the attributes of the individual impulses, but in the position of the fibre in the central organization of the nervous system.

[1] C. T. Morgan, 1943, p. 298; R. S. Woodworth, 1938, p. 465.
[2] E. G. Boring, 1942, p. 84.

4. DIFFERENCES IN QUALITY ARE DIFFERENCES IN THE EFFECTS

1.38. That the similarities and differences between the experienced sensory qualities do not correspond strictly to the differences and similarities between the physical attributes of the stimuli has become most familiar in connexion with the perception of configurations or gestalts. We all readily recognize as the same tune two different series of tones, or as the same shape or figure structures of different size and colour. In all these instances groups of stimuli which individually may be altogether different do yet as groups evoke the same sensory quality or are classified by our senses as the same gestalt.

1.39. But, though the fact that physically different stimuli produce similar sensory qualities is perhaps most conspicuous in connexion with the perception of 'wholes', it is no less present or less important where more simple or 'elementary' sensations are concerned. The fact that physically similar stimuli which act on different individual receptors and therefore set up impulses in different fibres evoke the same sensory quality raises a real problem. And the question why different physical stimuli for which different receptor organs are sensitive, and even physically similar stimuli acting on different kinds of receptor organs, should produce different sensations raises a problem of the same character.

1.40. While as a rule the same kind of physical stimuli acting on different receptor organs produce the same sensory quality, this is generally true only if they act on receptors of the same kind and even then not in all instances. The same vibration which, if perceived through the ear, will be experienced as a sound, may be experienced as a vibration by the sense of touch. In other instances 'the same external agent in one case produces light, in another warmth.'[1] The same temperature may be experienced as hot, cold, or pain according as it affects different end organs.[2] The same chemical stimulus may produce different sensory qualities according as it affects the mucous membranes of the eye or of the mouth.[3] And an electrical stimulation seems to be capable of evoking an even greater variety of different sensations. Moreover, even the same stimulus affecting the same receptors may produce

[1] E. Hering, 1885, (1913) p. 26.
[2] H. Head, 1920, II p. 807.
[3] R. W. Moncrieff, 1944, p. 32.

different sensations according as different other stimuli operate at the same time on other parts of the nervous system.

1.41. The same sensory quality, on the other hand, may be evoked by different physical stimuli. This happens not only where a particular receptor organ is excited by several different stimuli. In such a case any one of the different stimuli will, of course, evoke the same impulse. But impulses or groups of impulses set up in different fibres by different stimuli also often produce the same sensory quality. The classical instance is the case of colour vision and particularly the sensation of 'white' which can be produced by an infinite variety of different mixtures of light rays. But this same fact that physically different stimuli acting on different kinds of receptors produce the same sensory qualities seem to be of very frequent occurrence.

1.42. There exists, therefore, no one-to-one correspondence between the kinds (or the physical properties) of the different physical stimuli and the dimensions in which they can vary, on the one hand, and the different kinds of sensory qualities which they produce and their various dimensions, on the other. The manner in which the different physical stimuli can vary and the different physical dimensions in which they can be arranged have no exact counterpart in the manner in which the sensory qualities caused by them will differ from each other, or in the dimensions in which these sensory qualities can be arranged. This is the central fact to which we have referred when we insisted that the two orders, the physical order of the stimuli and the phenomenal or mental order of the sensory qualities, are different.

1.43. It has long been believed that, e.g., in the field of vision the three dimensions of the stimulus, wave-length, homogeneity and intensity correspond to the three phenomenal dimensions of visual experience, hue, saturation and brightness, and that similarly in the field of hearing frequency and intensity as physical dimensions of the stimulus correspond to pitch and loudness respectively as the phenomenal dimensions of sensation. Recent work, however, has amply shown that within any given modality a change in one dimension of the stimulus may affect almost any dimension of the sensation. Hue depends not only on wave-length but also on intensity; pitch not only on frequency but also on intensity.[1]

[1] See, e.g., S. S. Stevens, 1934; S. S. Stevens and H. Davis, 1938, p. 160; E. G. Boring, 1942, pp. 89, 376; F. L. Dimmick in Boring, Langfeld and Weld, 1948, pp. 270–280.

1.44. The orders or dimensions of the stimuli and of the sensations, moreover, not only show no one-to-one correspondence; they also differ in their general character. Any one of the physical dimensions of light and particularly wave-length which is mainly (though not exclusively) the cause of variation in colour, varies on a linear scale, while phenomenal colours can be arranged in a continuous circle in which the order of the wavelength is preserved, but the gap between the two extremes of the spectrum, yellowish red and violet, is closed by pure (or 'unique') red and purple which correspond to no distinct wave-length but can be produced only by various mixtures of different wave-lengths. Moreover, continuous variations of the stimuli often produce discontinuous variations in the sensory qualities[1], while in at least one case a continuous variation in the sensory qualities, namely from cold to hot, is brought about by what we must regard as a discontinuous variation of the stimuli, since the objectively continuous variation of temperature acts on the organism through different receptor organs.

1.45. It may be generally said that the organization of the sensory order, as represented by the various geometrical figures (such as the colour octohedron, Titchener's touch pyramid, Henning's smell prism and taste tetrahedron) by which psychologists have described the dimensions in which the sensory qualities vary, are by no means identical with the order of the corresponding physical stimuli and often differ very substantially from them. The fact that the two orders resemble each other in some degree must not obscure the fact that they are distinct and different orders.

1.46. When we speak of the physical order we mean by similarity of two events that they will produce the same effects in certain circumstances but not in others. Different physical events can evidently be similar to each other both in different degrees and also in different respects: two events may each be similar to a third but not be similar to each other. In other words, similarity is a non-transitive relation.

1.47. The same is true with regard to mental events. Two sensory qualities will be equal if their effects on other mental events or on behaviour will be the same in all respects. They may be similar in varying degrees and in different respects according as they will

[1] V. v. Weizsaecker, 1947 [1940], pp. 15–16.

evoke the same other mental events or the same behaviour in certain circumstances but not in others.

1.48. It will now be clearer what we mean when we speak of the two orders of events, the physical[1] and the phenomenal or mental order. Some events will occupy definite positions in both orders, but the relations between several such events in each of the two orders may be different. Some events in the physical order, such as electrical currents which we can only infer, will have no corresponding events in the phenomenal order; and some events in the phenomenal order, such as images or illusions which are not produced by external stimuli, will have no counterpart in the physical order. While there will thus be some degree of correspondence between the individual events which occur in the two orders, it will be but a very imperfect correspondence.

1.49. *What we call 'mind' is thus a particular order of a set of events taking place in some organism and in some manner related to but not identical with, the physical order of events in the environment.*[2] The problem which the existence of mental phenomena raises is therefore how in a part of the physical order (namely an organism) a sub-system can be formed which in some sense (yet to be more fully defined) may be said to reflect some features of the physical order as a whole, and which thereby enables the organism which contains such a partial reproduction of the environmental order to behave appropriately towards its surroundings. The problem arises as much from the fact that the order of this sub-system is in some respects similar to, as from the fact that it is in other respects different from the corresponding more comprehensive physical order. The meaning of the conception of an 'order' will be further explained in the next chapter (2.28–2.30).

1.50. In recent physiological psychology these problems have received attention, mainly owing to the work of H. Klüver, under the headings of *equivalence of stimuli* and of *sensory generalization*. Klüver's original statement of the problem is probably still the

[1]It is, perhaps, not inappropriate at this point explicitly to remind the reader that in this context 'physical order' refers exclusively to the order of the external stimuli and not to the order of the physiological impulses which, of course, also form part of the physical order in a wider sense. The nature of this order of the impulses will be considered in the next chapter.

[2]Cf., G. Ryle, 1949, p. 167: 'When we speak of a person's mind . . . [we are speaking of] . . . certain ways in which some of the incidents of his life are ordered.'

clearest exposition of it to be found in the literature.[1] Merely another aspect of the same problem is the phenomenon of *transfer* of acquired responses from a given stimulus to others, which is of course the process through which phenomenal similarity manifests itself in behaviour. Yet, though the central character of this problem is now fairly generally recognized, it is usually mentioned merely to point out that it is 'one of the most perplexing problems to be faced'[2] or 'the recognized stumbling block to all simple mechanical hypotheses of habit formation.'[3]

1.51. Equivalence, generalization and transfer are all instances of identity of the effects of different stimuli, while discrimination means a difference in the effect of individual stimuli or groups of stimuli. The qualitative order of the sensations which manifests itself in these phenomena is thus a difference in the order in which the stimuli in various combinations produce different effects; and sensory qualities can be regarded as groups or classes of events which, with respect to the responses of the organism, are identical, similar or different in their effects. The order of sensory qualities thus is identical with the totality of the differences of the effects which the different nervous impulses will produce in different circumstances. If we can explain the process which determines the differential responses of the organism to the various physical stimuli, we have at the same time also explained the qualitative order which is the peculiar characteristic of mental phenomena.

1.52. The significance of this statement, which in its bare form may sound more 'behaviouristic' than it is intended, will become clearer when we examine the kinds of different 'effects' which have to be considered in this connexion (2.23–2.26). At this point it need only be pointed out that by the term 'effects' we do not mean only, or even mainly, overt behaviour or peripheral responses, but shall include all the central nervous processes caused by the

---

[1]H. Klüver, 1933, especially pp. 330–332; 1935, p. 109; and 1949, p. 404. A clear statement is also to be found in E. R. Hilgard and D. G. Marquis, 1940, p. 176: 'The basic facts of stimulus equivalence and response equivalence are not limited in application to conditioned responses, but are true of reflexes and of complex voluntary responses. Every response is elicitable not just by one stimulus but by a *class* of stimuli. Correspondingly, every stimulus elicits, not just one response, but one of a *class* of responses.' (Italics ours).
[2]C. T. Morgan, 1943, p. 514.
[3]E. D. Adrian, 1947, p. 82.

initial impulses, even though we may be able only indirectly to infer their existence.

1.53. Our problem is then to show how it is possible to build from the known elements of the nervous processes a structure of intermediate links between the physical stimuli and the overt responses which can account for the fact that the responses to different stimuli differ from each other in precisely that fashion in which we know the responses to the experienced sensory qualities to differ from each other. We must show that from the known physiological elements a structure can be formed which can differentiate between different impulses passing through it in exactly the same manner in which our sensory experience differentiates between the different stimuli.

1.54. Our problem must therefore be *stated* in terms of the relationships (of equality, similarity, difference, etc.) existing between the sensory qualities. It can be *answered* only by showing that a strictly equivalent system of relationships can exist between physiological events so that the effects of any event or any group of events in that system will produce a set of effects strictly corresponding to the effects the corresponding sensory qualities will produce. (The reader should observe already at this stage that this does not imply that any given physiological event will always produce the same effects irrespective of the other physiological events occurring at the same time. On this and on the general danger of a too narrow interpretation of the conception of a one-to-one correspondence between the sensory and the neural order see below 2.10–2.13.)

1.55. This contention implies that if we can explain how all the different sensory qualities differ from each other in the effects which they will produce whenever they occur, we have explained all there is to explain; or that the whole order of sensory qualities can be exhaustively described in terms of (or 'consists of nothing but') all the relationships existing between them.[1] There is no problem of sensory qualities beyond the problem of how the

---

[1]That this is the consistent development of the approach started by John Locke was clearly seen by T. H. Green who argued (1884, p. 23) that 'if we take him [Locke] at his word and exclude from what we have considered real all qualities constituted by relation, we find that none are left. Without relation any simple idea would be undistinguished from other simple ideas, undermined by its surroundings in the cosmos of existence.' See also *ibid.*, p. 31.

different qualities differ from each other—and these differences can only consist of differences in the effects which they exercise in evoking other qualities, or in determining behaviour.

5. THE UNITARY CHARACTER OF THE SENSORY ORDER

1.56. The conclusion to which we have been led means that the order of sensory qualities no less than the order of physical events is a relational order—even though to us, whose mind is the totality of the relations constituting that order, it may not appear as such. The difference between the physical order of events and the phenomenal order in which we perceive the same events is thus not that only the former is purely relational, but that the relations existing between corresponding events and groups of events in the two orders will be different.

1.57. The order of the sensory qualities is difficult to describe, not only because we are not explicitly aware of the relations between the different qualities but merely manifest these relations in the discriminations which we perform,[1] and because the number and complexity of these relations is probably greater than anything which we could ever explicitly state or exhaustively describe, but also because, as we shall see, it is not a stable but a variable order. Yet we must attempt here to describe at least certain general characteristics of that order, because our problem is whether we can account for at least the kind of properties which it possesses, even if we cannot explain its detailed arrangement.

1.58. One main point about this order is that, in spite of its division into the different modalities, it is still a unitary order, in the sense that any two events belonging to it may in certain definite ways resemble each other or differ from each other. Any colour and any smell, any tone and any temperature, or any tactual sensation such as smoothness or wetness and any experience of shape or rhythm may yet have something in common, or be at least in some sense akin to or in contrast with one another. Experiments have shown that these experienced similarities extend much further than we are usually aware of and that, e.g., even a

---

[1]This distinction is probably the same as, or closely related to, that between 'Knowing How' and 'Knowing That' so well brought out by G. Ryle, 1945 and 1949.

person who at first thinks such an attempt nonsense, has no difficulty, once he can bring himself to try, to find a tone whose brightness is the same as that of the smell of lilac.[1]

1.59. Some qualities, especially those which, like colours or tones, are connected into qualitative continua and which, since Helmholtz, we describe as forming distinct modalities, probably always seem to belong more closely together than others such as, e.g., the sensations of pressure, pain, and temperature, which used to be regarded as belonging to the one sense of touch but do not form one modality in the sense just defined. But when we try to describe the differences between different qualities belonging to the same modality, such as different colours, we find that in order to do so we usually resort to expressions borrowed from other modalities. One colour may be warmer or heavier or louder than another, one tone brighter or rougher or thicker than another. This indicates that, though in some respects one particular colour or one particular tone may be most closely related to other colours or other tones respectively, yet in other respects they may be closer to qualities belonging to different modalities.

1.60. Although within any given modality qualities vary continuously[2] they need not vary in a constant direction or dimension. While it is true of tones that if one tone is higher than a second, and a third higher than the first, the third will also be higher than the second, we cannot similarly say that, because orange is yellower than red and green bluer than orange, green is therefore either more yellow or more blue than red. While, with regard to pitch, tones can be arranged in one linear scale, colours do not, in this sense, vary in a single direction.

1.61. It makes sense, on the other hand, to say that two different colours differ in the same manner in which two different temperatures or weights do, or that two tones differ similarly as do two sensations of colour or touch. This means that qualities of different modalities may vary along similar or parallel directions or dimensions, or that the same kind of differences can occur in different modalities. It is, e.g., part of the difference between blue and red that blue is associated with coolness and red with warmth.

[1]E. M. von Hornbostel, 1925, p. 290.
[2]Some doubt has recently been thrown even on the complete continuity of the qualities within one modality and the existence of sensory 'quanta' suggested by S. S. Stevens and J. Volkman, 1940 and 1941.

There exist apparently certain intermodal or intersensory attributes, and with regard to some of the terms which we use for them, such as strong or weak, mild or mellow, tingling or sharp, we are often not immediately aware to which sense modality they originally belong.[1]

1.62. In our highly developed conscious picture of the sensory order these intersensory and intermodal relations are not very prominent and with the development of conceptual thought and particularly, as a result of the great influence which sensualism has had on it, in scientific thought, they are more and more driven back until they are almost completely disregarded.[2] We may become aware of their existence only when we attempt to describe a particular sensory quality and in doing so find ourselves driven to describe a colour as soft or sweet, a tone as thin or dark, a taste as hot or sharp, or a smell as dry and sweet. There can be little doubt that these seemingly metaphorical expressions refer to truly intersensory attributes; and experimental tests have at least in some instances shown that different people tend to equate the same pairs or groups of different qualities.[3]

1.63. These facts may also be described by saying that relations between different qualities may in turn also possess distinct qualities, and that the relations between different pairs or groups of qualities belonging to different modalities may possess the same qualities. These qualities attaching to the relations between different qualities may in turn be similar to individual sensory qualities. The successive musical intervals from the second to the octave, e.g., have been described as 'gritty', 'mellow', 'coarse',

---

[1]See especially M. Schiller, 1932, and the instances of terms borrowed from other modalities to describe smells given by F. W. Hazzard, 1930, p. 318. It is also interesting to note that the meaning of the German word *hell* (bright) has shifted from its original reference to auditory experience to the visual field.

[2]Very characteristic in this connexion is the categorical statement by M. Planck, 1949 (1941), p. 87, that the experiences of the different sensory fields 'are totally different from each other, and have initially nothing in common. There is no immediate, direct bridge between the perception of colours and the perception of sounds. An affinity, such as may be assumed by many art lovers to exist between a certain shade of colour and a certain musical pitch, is not directly given but is the creation, stimulated by personal experiences, of our reflective powers of imagination.' The fact seems to be the other way round that sophistication makes us overlook what is obvious to naïve experience.

[3]On this and the following see G. M. Hartmann, 1935, pp. 141–151.

'hollow', 'luscious', 'astringent', and 'smooth' respectively.[1]

1.64. These intermodal relations may occasionally be so strong that different sensations belonging to one modality may regularly be accompanied by the experience of qualities belonging to another modality, as in the case of colour-hearing and other instances of synaesthesia. There is some evidence that these synaesthetic modes of perception are particularly strong in relatively early stages of mental development, and that our habit of thinking of particular colours as primarily belonging to the range of colours, or of a tone primarily as being one of a range of tones, is the produce of a comparatively late and abstract attitude.[2]

1.65. More familiar than the facts of synaesthesia is the fact that most sensory qualities are closely associated with certain affective tones and that there exists thus a close connexion between the order of sensory qualities and that of affective qualities. The emotional values attaching to various sensory qualities are well known, and there are indeed few sensory qualities which we do not regard at least as either pleasant or unpleasant, or as simply good or bad. The general relation between sensations and emotions or drives will, however, have to be considered later and cannot be further examined at this point.

1.66. The relations or connexions between different sensory (and affective) qualities find expression in the expectations which their occurrence arouses. A red colour does not merely evoke the image of warmth but we shall be rather surprised if a red objects turns out to be very cold; and a certain smell will not only conjure up certain tastes but we shall be shocked if a deliciously smelling fruit turns out to have a vile taste. In this way certain groups of qualities tend to 'belong' together, and particular qualities come to 'mean' to us certain other qualities.

1.67. Whether the facts briefly summarized in this section do or do not justify the assertion of a 'Unity of the Senses' in such a manner that 'all senses are alike in respect to their attributive dimensions',[3] they probably entitle us to say that, directly or

---

[1] E. M. Edmonds and M. E. Smith, 1923.
[2] H. Werner, 1948, p. 86. On Synaesthesia see also H. Kleint, 1940, pp. 56–61, K. Goldstein, 1939, p. 267 and for bibliographies of the extensive literature on the subject, F. Mahling, 1926, A. Argelander, 1927, and A. Wellek, 1931.
[3] This is the interpretation given to E. M. von Hornbostel's conception of the *Unity of the Senses* by E. G. Boring, 1942, p. 27.

indirectly, all mental qualities are so related to each other that any attempt to give an exhaustive description of any one of them would make it necessary to describe the relations existing between all.

6. THE ORDER OF SENSORY QUALITIES NOT CONFINED TO CONSCIOUS EXPERIENCE

1.68. We have so far assumed that the reader is familiar with the system of sensory qualities from his own conscious experience of these qualities. This, however, is not to be understood to mean that this particular classification of events appears only in our subjective experience. Of course we know this system of qualities from this source. But just as experience tells us that in their relations to each other things do not always resemble each other or differ from each other in the same manner as they seem to be alike or different to us, so we also learn that what appears alike or different to us usually also appears alike or different to other men. Beyond this, it seems clear that not only other men in their conscious action, but both we and others in unconscious action, and also animals, treat as alike or different not what is so in the physical sense, but more or less what in our own conscious experience appears to us to be so. In other words, the order of sensory qualities, once it is known, can be recognized as present in actions which are not directed by consciousness or by a human mind.

1.69. It would, of course, not be possible to discuss the phenomenal world with other people if they did not perceive this world in terms of the same, or at least of a very similar, order of qualities as we do. This means that the conscious mind of other people classifies stimuli in a manner similar to that in which our own mind does so, and that the different sensory qualities are for them related to each other in a manner which is similar to that which we know. In other words, although the system of sensory qualities is 'subjective' in the sense of belonging to the perceiving subject as distinguished from 'objective' (belonging to the perceived objects)— a distinction which is the same as that between the phenomenal and the physical order—it is yet inter-personal and not (or at least not entirely) peculiar to the individual.

1.70. Nor is the classification of stimuli in terms of sensory qualities confined to conscious experience. We know that both we and other people classify stimuli in our unconscious responses (or in responses to stimuli of which we do not become conscious) according to roughly the same principles as we do in our conscious action.[1] The order of sensory qualities exists therefore also outside the realm of consciousness. If, as we shall suggest, we identify with the realm of mental phenomena the range of events within which a classification in terms of sensory (and similar mental) qualities occurs, this realm extends far beyond the sphere of conscious events which merely constitute a special group within the more comprehensive class of mental events.

1.71. It is possible, finally, to ascertain by various experimental methods that not only other men but also most higher animals classify stimuli according to an order which is similar to that of our own sensory experiences. It has even been shown that some animals, e.g., chicks in the famous Révész experiment,[2] are subject to the same optical illusions as men. We must therefore conclude that the general principles according to which the neural system of the higher animals classifies stimuli are, at least in their general outline, similar to those on which our own mind operates.

1.72. While it has been inevitable that in introducing our problem we started from the conscious experience of sensory qualities, this proves now to be only one particular aspect of a wider problem. In the further discussion we shall treat conscious experience as merely a special instance of a more general phenomenon, and speak of mental phenomena whenever we deal with any events which are ordered on principles analogous to those revealed by conscious experience. All further consideration of the peculiar additional attributes which a mental event in this sense must possess in order to be described as 'conscious' will be postponed to a later stage (Chapter VI).

1.73. It has undoubtedly been unfortunate for the development of psychology that the distinguishing attribute of its object was so long considered to be the 'conscious' character of experience, and that no definition of mental events was available which was

---

[1]For the fact that this applies even to responses to configurations see K. Lorenz, 1943, p. 323; and on subconscious discrimination ('subception') R. A. McCleary and R. S. Lazarus, 1949, p. 178.
[2]G. Révész, 1924, and C. N. Winslow, 1933.

independent of this conscious character.[1] The sphere of mental events evidently transcends the sphere of conscious events and there is no justification for the attitude frequently met that either identifies the two or even maintains that to speak of unconscious mental events is a contradiction in terms.[2]

1.74. But although we can agree with the Behaviourists in deploring the exclusive concentration of the older psychology on conscious events, they themselves, in their endeavour to get rid of consciousness have gone to the opposite extreme and with the problem of consciousness have tried to eliminate the problem of the existence of the qualitative order which is peculiar to mental phenomena. This problem, as we shall see, cannot be disregarded even if we want merely to account for observed behaviour.

7. THE DENIAL OR DISREGARD OF OUR PROBLEM BY BEHAVIOURISM

1.75. It will help to bring out more clearly the precise meaning of our problem if we contrast our approach with that of two other points of view which require either less or more of any explanation of sensory perception than our statement of the problem demands. This and the next section will accordingly be devoted to an examination, firstly, of the views of a school of thought which either explicitly denied the existence of our problem, or at least proceeded as if it did not exist; and, secondly, to the consideration of

[1]Cf., E. B. Holt, 1937, p. 41: 'Every school of psychology since certainly before the time of Herbart has found that by far the greater portion of the sensations, ideas and processes which must be called "mental" never become explicitly conscious: they are not perceived and cannot by any known process of introspection be perceived.' Also the passage quoted by Holt from S. Freud, 1918, p. 9, where the latter says that 'mental processes in and of themselves are unconscious and the conscious are merely isolated acts and passages in the total life of the mind.' Cf., also E. G. Boring, 1948, on the use of the term 'unconscious mind.'

[2]Several examples of the identification of 'mental' and 'conscious' are given by J. G. Miller, 1942, pp. 24ff. C. J. Herrick, 1926, p. 280 says that 'the dynamic view of consciousness here adopted makes such expressions as "the unconscious mind" impossible contradictions.' H. Head, 1920, II, p. 747, states that 'sensation, in the strict sense of the term, demands the existence of consciousness.' M. Planck, 1949, p. 66 also describes a 'science of the unconscious or subconscious mind' as 'a contradiction in terms, a self-contradiction.'

an opposite point of view which would probably maintain that even if a complete answer to our problem were achieved, there would still remain unsolved a significant problem concerning the 'absolute' or 'intrinsic' nature of sensory qualities.

1.76. The point of view which denies, at least by implication, that ours is a genuine problem is (or was?) represented mainly by the classical behaviourists[1] and by similar schools aiming at a strictly 'objective' psychology. These schools maintained that psychology can entirely dispense with any knowledge of the subjectively experienced mental qualities, and that it ought to confine itself to the study of bodily responses to physical stimuli.

1.77. All the schools of psychology which thus claim to confine themselves to observed physical facts, are, however, in fact, always and inevitably inconsistent in their procedure: they never really avoid using knowledge which according to their professed principles they have no right to use. They almost invariably describe the external stimuli which elicit behaviour not in terms of their physical properties but in terms of their sensory attributes. They naïvely accept as a fact not requiring explanation that different minds treat as equal, similar, or different, groups of stimuli, which physically are not such but merely appear so to our senses.

1.78. The adherents of these schools, in other words, treat as something not requiring explanation the fact that stimuli, which to their senses appear similar, will also appear so to others; and they do this in spite of our knowledge that physically these stimuli may be very different events and in fact may have nothing in common except that very circumstance that whenever they act on us or other people they will evoke the same sensations (and/or responses). They disregard, in other words, the very phenomenon

---

[1] By 'Behaviourism' we shall mean throughout this discussion not only the original doctrines of J. Watson but also the views represented in the nineteen-twenties and early 'thirties by men like E. B. Holt, A. P. Weiss, E. C. Tolman, W. S. Hunter and particularly K. S. Lashley, who in 1923 defined the position by the statement that 'the behaviourist denies sensations, images, and all other phenomena which the subjectivist claims to find by introspection.' More recently this radically objectivist attitude has been greatly modified and one may doubt whether the Lashley who (1942, p. 304) has 'come to doubt that any progress will be made towards a genuine understanding of nervous integration until the problem of equivalent nervous connexions, as it is more generally termed, of stimulus equivalence, is solved,' can still be described as a behaviourist. See also K. W. Spence, 1948, p. 67.

which raises the problem of the existence of a peculiar mental order.

1.79. It might therefore be said that behaviourism, from its own point of view, was not radical and consistent enough, since it took for its starting point a picture of the external world which was derived from our naïve sense experience, instead of taking, as it ought to have done, one obtained from the physical sciences which describe the objective properties of this world. If the behaviourists had been consistent in their desire to take no notice of the qualitative order of their own sense experience, they ought to have started by studying the effects on the organism of physical events of a certain kind, e.g., of light waves of a certain frequency, and then have proceeded to establish experimentally to which of these different physical stimuli the individual responded in the same and to which he responded in a different manner. Before going any further they ought, in other words, to have built up experimentally that classification of the different stimuli which our senses effect.[1]

1.80. Behaviourists, however, did not seriously try doing anything of the kind. They uncritically accepted the fact that things which are physically different appear alike to our senses, and that things which are physically the same sometimes appear different, or that different things may appear to differ from each other in a manner which is in no way commensurable with the physical differences which objectively exist between them; and they appeared to see no problem in the fact that other organisms classify stimuli in the same manner as we do ourselves, or in a manner different from it.

1.81. This curious blindness to an important problem does not always show itself as blatantly as in the instance reported by W. Köhler in which a behaviourist insisted on referring to a 'female' as 'a stimulus' to a male bird.[2] The error in this instance does not lie merely, as Köhler suggests, in the fact that it involves 'closing one's eye to the problem of gestalt and organization.' It appears already in the disregard of the fact that physically different stimuli affecting different receptors produce the same or similar sensory qualities and therefore are treated as being the same, and in

[1] Cf., F. A. Hayek, 1943, pp. 34–39.
[2] W. Köhler, 1929 p. 180: 66. C.f., Also E. G. Boring, 1930, p. 121: 'Green light of 505 millimicrons wave-length may be a stimulus but my grandmother is not a stimulus', and W. Metzger, 1941, p. 283.

pretending at the same time that sensory qualities do not enter at all into their considerations. (The language of the behaviourist in this instance could be justified only if he meant to imply that the female was always recognized through the same physical stimulus, such as a certain smell, or rather by the stimulation of certain organs of olfaction by definite chemical substances.)

1.82. It would involve the same disregard of the central problem if, e.g., two red spots reflected on different parts of the retina, or the same temperature affecting different parts of the body,were treated as representing the same stimulus. In treating as the same kind of event all events which appear to us to possess the same sensory qualities, behaviourism tacitly assumes the existence of the whole order of such qualities which at the same time it pretends to ignore.

1.83. This acceptance as data of the sensory qualities as they are known to most men from their subjective experience is indeed inevitable in the study of any complex behaviour. But it is only because, while thus accepting them, the behaviourists at the same time deceived themselves about the true character of their procedure, that they avoided the main problem which psychology has to face. If they had been more radical and more consistent in their efforts to link up psychology with the world of physical science, they would have discovered[1] that their attempt to explain behaviour without reference to subjective sensory qualities could not be consistently carried through unless it was first shown what determined that system of sensory qualities.

1.84. Like many of the traditional schools of psychology, behaviourism thus treated the problem of mind as if it were a problem of the responses of the individual to an independently or objectively given phenomenal world; while in fact it is the existence of a phenomenal world which is different from the physical world which constitutes the main problem. Behaviourism merely appeared to avoid the problem of mind by confining itself to the study of man's behaviour in the phenomenal world and by thus treating the main manifestation of mind as a datum rather than as something requiring explanation.

[1] As they ultimately did—see the passage from Lashley quoted to 1.76 above. One might indeed date the end of behaviourism at the time of a general recognition of the central importance of the problem of equivalence of stimuli, i.e., soon after the appearance of H. Klüver, 1933.

1.85. Although no behaviourist ever consistently adhered to what are the professed principles of his school, and although, if he had, he would never, in the present state of knowledge, have got on to the phenomena in which he was interested, it will be instructive briefly to consider what a consistently 'objectivist' study of behaviour would have to be like. It will then be seen that even if the behaviourists had succeeded in carrying out their programme, there would still remain a problem of mind requiring an answer.

1.86. In the first instance, much knowledge that we undoubtedly possess but which is not derived from experimental evidence—such as the knowledge that we are likely to respond in the same manner to different physical stimuli which produce the same sensation—would have to be strictly excluded from such a study of human behaviour. The first task of such a consistently objectivist approach would therefore have to be to ascertain experimentally what to us is the starting point of all knowledge, namely the phenomenal order in which the different stimuli appear in our mind.

1.87. It is at least not inconceivable, although not likely, that by proceeding thus we might in the course of time succeed in reconstructing approximately that grouping of the stimuli which our senses perform. We might then be able to list all the different physical stimuli which, acting on particular receptors and under particular conditions, produce the same sensations (or have always the same influence on the response), and also to reconstruct all the different conditions under which (and all the different respects with regard to which) the several stimuli produce different effects. In other words we might, starting from the physical order of events, experimentally reconstruct the phenomenal order in which these events are reproduced by our senses.[1]

1.88. This would be merely the first task which a psychology would have to undertake which took the basic idea of behaviourism literally. Only after completing this task could it at least undertake to link directly observable behaviour and physical stimuli.

---

[1]This would require more than that co-ordination of the dimensions of individual stimuli and the dimensions of the various 'elementary' sensory qualities which recently has been successfully attempted, especially with regard to hearing, by S. S. Stevens, 1934. It would require a similar co-ordination also for the instances where the same stimulus in different combinations with others produces different sensations.

And in order to be quite consistent it would have to define not only the stimuli but also behaviour in strictly physical terms. We need not inquire at this stage whether it is conceivable that this task should ever be fully completed. (We shall later give reasons why we think that this is impossible.) At this point we are concerned with the question whether, even if this task were achieved, there would still remain a problem of the kind with which we are here concerned.

1.89. A solution of that problem would show us what the apparatus of perception does in response to particular stimuli, but not how it does it. Even if we had established a correspondence between all the observed combinations of stimuli and the resulting sensations, we should still be ignorant of the mechanism by which the one kind of order is translated into the other. Our knowledge would be purely descriptive in the sense that it would be confined to a knowledge of the correspondence between observed stimuli and observed responses. We should not possess a theory from which we could derive new conclusions which could be empirically tested.

8. THE 'ABSOLUTE' QUALITIES OF SENSATIONS A PHANTOM-PROBLEM

1.90. A different type of objection to our manner of stating the problem must be expected from a school of thought which, though not formally organized, is fairly widespread and which in some respects might be regarded as the extreme opposite of behaviourism. It would probably be contended by representatives of this point of view that, even if we succeeded in accounting for all the differences between the effects of the different stimuli or impulses, there would still remain an unexplained factor, the 'absolute' or 'intrinsic' qualities of the sensations which are not exhausted by all the differences in their effects but which must be experienced to be known.

1.91. This conception of the absolute character of sensory qualities derives probably from John Locke's conception of 'simple' ideas. It has found an explicit defender in no less a student than William James.[1] It is a contention which raises what to us seems clearly a phantom-problem which cannot even be clearly stated

[1]W. James, 1890, II, p. 12.

and with regard to which it is impossible to say what kind of statement would provide an answer. It is nevertheless important, not only because of the pervasive influence of this conception, but also because it is probably one of the main roots of the belief in a peculiar mental substance.

1.92. The first point to note is that it is clearly possible that a sense discrimination of which some other person is capable can raise a problem for us though we ourselves may not be capable of it. The problem of colour vision, e.g., can clearly become a problem to the totally colour-blind person as much as it can to us. What we shall have to show is that there are no questions which we can intelligibly ask about sensory qualities which could not also conceivably become a problem to a person who has not himself experienced the particular qualities but knows of them only from the descriptions given to him by others. In other words, that nothing can become a problem about sensory qualities which cannot in principle also be described in words; and such a description in words will always have to be a description in terms of the relation of the quality in question to other sensory qualities.

1.93. Most people will agree that the question of whether the sensory qualities which one person experiences are exactly the same as those which another person experiences is, in the absolute sense in which it is sometimes asked, an unanswerable and strictly meaningless question. All we can ever discuss is whether for different persons different sensory qualities differ in the same way. To establish whether a person is colour-blind we have to find out, not how 'red' looks to him in any absolute sense, but whether and how it differs from various other shades of 'red' and from 'green.' In all such instances we can find out and know only whether, compared with other people, a person discriminates between given stimuli in the same or in a different manner.

1.94. In other words, all that can be communicated are the differences between sensory qualities, and only what can be communicated can be discussed. Such communication does not imply that the qualities perceived by different people are similar in any absolute sense. The problem which is raised, for instance, by the much greater capacity for pitch discrimination possessed by the experienced musician but not by ordinary persons is not fundamentally different from the problem created by the distinctions between the qualities which most of us experience.

1.95. It is instructive briefly to consider how we should proceed if we were to try to give a congenitally blind person an idea of sight and colour. We should probably base our account in the first instance on the fact that the blind is familiar with three-dimensional space, with shape and movement, and attempt to explain to him that, as he can feel radiant heat or sound emitted by a distant source, so the eye enables us to perceive other qualities at a distance. We should then try to explain that these qualities with which he is unfamiliar will vary not only along a single dimension, as temperature does from cold to hot, but that it can also vary like tones from bright to dark, from loud to soft, from sharp to blunt and from pleasant to unpleasant. We shall point out to him that in groups these qualities can form harmonies or may clash as tones do, and so on.

1.96. How far we could get in thus teaching a congenitally blind the relative values of the different colours has never been systematically tested, largely because the required description of the order of those sense qualities in terms of their common dimensions (1.62–1.67) has not been systematically developed and because we therefore lack the necessary words. That blind persons can at least learn to use the names of colours so that a person who does not know that they are blind may remain unaware of it in hearing their descriptions is shown by the writings of Miss Helen Keller and others. To-day, with our greater familiarity of the phenomenon of synaesthesia, it also no longer seems so absurd, as it seemed to John Locke, that the 'studious blind man' who thought that he had discovered what scarlet looked like, described it as 'like the sound of a trumpet.'[1]

1.97. An illustration given in a recent book may be quoted at length, as its concluding passage raises our problem in a particularly clear manner:

'The approach of a scientist to the phenomena which he observes may be realized perhaps by means of an analogy. Suppose you enter a room and see a man playing a violin. You say at once that this is a musical instrument and is producing sound. But suppose that the observer were absolutely deaf from birth, had no idea of hearing, and had never been told anything of sound or musical instruments, his whole knowledge of the world having been acquired through senses other than hearing. This deaf

[1]John Locke, 1690, Bk. III, Chap IV, Sec. xi.

observer entering the room where a violinist was playing would be entirely unable to account for the phenomenon. He would see the movements of the player, the operation of the bow on the strings, the peculiarly shaped instrument, but the whole thing would appear to him irrational. But if he were a scientist interested in phenomena and their classification, he would presently find that the movements of the bow on the violin produced vibrations, and these vibrations could be detected by means of physical instruments and their wave form could be observed. After some time, it might occur to him that the vibrations of the strings and violin must be communicated to the air and could be observed as changes of pressure. Then he could record the changes of pressure produced in the air in the playing of a piece of music, and by analysing the record could observe that the same groups of pressure changes were repeated periodically. Eventually he would attain to a knowledge of the whole phenomenon of music—the form of musical composition and the nature of different musical forms— but none of this would give him any approach to the absolute truth in that he would still be unaware of the existence of sound as a sense and of the part that music could play in the mental life of those who could hear.'[1]

1.97. Except for the last sentence this passage provides an excellent illustration of the distinction we have drawn between the physical and the phenomenal order of events. The last sentence, however, raises two difficulties (apart from the fact that the author speaks of the 'phenomenon' of music where he refers to what we would describe as its physical equivalent). In the first instance the impression which this sentence conveys, that a 'knowledge of the whole phenomenon of music' can be attained without at the same time attaining some knowledge not only of the physical but also of the sensory attributes of these events is somewhat misleading. A reconstruction of the theory of music in the manner suggested would involve a study not only of the 'objective' attributes of sound but also a study of the manner in which the people producing the music deal with it. It would, e.g., have to include the discovery that for the musicians the continuum of sound waves of different frequencies was divided into discrete steps, so that all the waves belonging to certain narrow intervals were treated as alike or indistinguishable, while wave-lengths of intermediate intervals would not be employed at all; further, that of the distinct musical

[1]C. E. K. Mees, 1947, p. 59

notes thus determined some were treated as resembling each other and some as being related in other ways, that certain combinations of notes were preferred to others, and that certain successions of notes were in some respects treated as equivalents of other such successions, etc., etc.

1.99. The theory of music thus constructed would therefore not really refer to the relations between physical events or to relations between them defined according to the similarity or difference of their action on other physical events, but to elements defined in terms of their similarity or dissimilarity to the persons who wrote, played, or heard the music. It would be a theory, not about the objective (experimentally tested) relations between the different physical events, but about what these events meant to the persons concerned with music.

1.100. The second problem arising from the concluding sentence of the passage quoted is contained in the suggestion that there is an 'absolute truth', an absolute quality of sound as a sensory experience, which must forever remain inaccessible to the deaf from birth. The term 'absolute' used in this connexion unquestionably refers to some significant aspects of sensory experience. What we are denying is not that sensory qualities may possess attributes which those who cannot hear cannot learn about, but that whatever incommunicable attributes sensory qualities may possess can ever raise a scientific problem.

1.101. One fact which is probably referred to by the use of the term 'absolute' in this connexion is that, however far we may go in describing or explaining differences between sensory qualities, there will always remain some further differences which have not yet been enumerated. This is closely connected with a circumstance which we shall have to consider later, namely that, because of constitutional limitations of our mind, we shall never be able to achieve more than an explanation of the principle on which mind operates, and shall never succeed in fully explaining any particular mental act. But the fact that the differences between the different sensory qualities are too numerous and varied for us ever to be able to state them all, does not mean that any one of these differences should not be capable of becoming a problem to which, at least in principle, we may provide an answer.

1.102. It is merely another aspect of the same problem if it is pointed out that the immediate experience of a group of sensory qualities (say a number of sounds and colours) will always convey more to us (will involve a large number of implied distinctions among themselves and from other possible experiences) than any possible description can convey. In other words: the congenitally blind or deaf can never learn *all* that which the seeing or hearing person owes to the direct experience of the sensory qualities in question, because no description can exhaust all the distinctions which are experienced. This, however, does not mean that there is more than differences from other qualities, and still less that any such 'absolute' character of the qualities can raise a genuine problem.

1.103. It seems thus impossible that any question about the nature or character of particular sensory qualities should ever arise which is not a question about the differences from (or the relations to) other sensory qualities; and the extent to which the effects of its occurrence differ from the effects of the occurrence of any other qualities determines the whole of its character.

1.104. To ask beyond this for the explanation of some absolute attribute of sensory qualities seems to be to ask for something which by definition cannot manifest itself in any differences in the consequences which will follow because this rather than any other quality has occurred. Such a factor, however, could by definition not be of relevance to any scientific problem. The 'absolute' quality seems to be unexplainable because there is nothing to explain, because absolute, if it has any meaning at all, can only mean that the attribute which is so described has no scientific significance.

1.105. The contention that all the attributes of sensory qualities (and of other mental qualities) are relations to other such qualities, and that the totality of all these relations between mental qualities exhausts all there is to be said about the mental order, corresponds of course, (perhaps we should say follows from) the conception of mind itself as an order of events. And with the recognition that mind itself, and all the attributes of mental events, are a complex of relations, there disappears of course the need for any peculiar kind of things which by themselves have attributes which constitute them a peculiar 'substance'.

1.106. The abandonment of the phantom-problem of the absolute character of mental qualities, and the recognition of the relative significance of these attributes, is of fundamental importance, because it opens, as we shall see, the way for a general application of a principle which has long been used to explain those attributes of sensory experience which had been recognized to be relative, such as spatial position.

1.107. It also follows from the relative character of all mental qualities that any discussion of these qualities in terms of their relations to each other must necessarily remain within the realm of mental events: it can never provide a bridge which leads from them to physical events. In the next chapter we shall attempt to show how this circle can be broken.

CHAPTER II

AN OUTLINE OF THE THEORY

I. THE PRINCIPLE OF THE EXPLANATION

2.1.   The first chapter led to the conclusion that the sensory qualities known to us from our subjective experience form a self-contained system so that we can describe any one of these qualities only in terms of its relations to other such qualities, and that many of these relations themselves also belong to the qualitative order. This means that, if in our attempt toward an explanation we are not to move in a circle but are to succeed in explaining the relation of this system of qualities to the world of physics, the object of our explanation must be the whole complex of relations which determine the order of the system of sensory (or rather of mental) qualities. In order to provide such an explanation, it will be necessary to show how in a physical system known forces can produce such differentiating relationships between its elements that an order will appear which strictly corresponds to the order of the sensory qualities.

2.2.   The only way in which we can break the circle in which we move so long as we discuss sensory qualities in terms of each other, and can hope to arrive at an explanation of the processes of which the occurrence of sensory qualities forms a part, therefore, is to construct a system of physical elements which is 'topologically equivalent' or 'isomorphous' with the system of sensory qualities; this means that the relations of the former must strictly reproduce the relations prevailing in the latter so that the effect of any groups of events in the former will correspond to the effects of the corresponding group of events in the latter.

2.3.   The mathematical concept of isomorphism has been used by the members of the gestalt school[1] in a sense somewhat similar to

---

[1] W. Köhler, 1929 p. 61 f., K. Koffka, 1935, p. 62. For the different sense in which this concept is used by E. G. Boring, see below, 2.10.

that in which it is employed here. The use made of it by that school is, however, somewhat ambiguous and imprecise and I am not certain whether it is the same as that employed here. It is therefore important to remember that, whenever the term isomorphism is used in the following discussion, it will be used in its strict mathematical meaning of a structural correspondence between systems of related elements in which the relations connecting these elements possess the same formal properties, rather than in any sense borrowed directly from the gestalt school.

2.4. It is especially important to realize that the isomorphism of two structures does not, as some of the discussions by the gestalt school suggest, imply similarity of their arrangement in space. Although two three-dimensional structures which are similar in the geometrical meaning of this term will also be isomorphous, such spatial similarity is not necessary. If the relevant relationship is, e.g., connectedness, and we conceive of a three dimensional net or lattice of rubber threads in which the knots represent the elements and the threads the connexions, isomorphism will be preserved however much we stretch, twist or crumple up the net. Since in this process of spatial distortion the relevant relations between the elements are preserved so long as no thread is broken and no new knots formed, all these various states of the net or lattice would be isomorphous. It will have to be remembered throughout this book that whenever we speak, e.g., of a 'pattern within the brain', the term pattern and similar terms will have to be understood in this topological and not in a spatial meaning.

2.5. The importance of not interpreting isomorphism as spatial similiarity will be seen from the fact, for instance, that in a system in which the position of one element is determined by the connexions with other elements, two distinct elements may occupy identical positions, which is clearly impossible in a spatial sense. Two distinct points in space cannot have identical spatial relations to every one of a group of other points, but each of two distinct elements of a merely 'connexional' order can be connected with the identical set of other elements. This in fact applies not only to individual elements but also to subgroups of connected elements within the larger structure which, without being isomorphous with each other, may yet, considered as groups, occupy identical places in the larger structure, i.e., may as a group of elements have connexions with the same other elements.

2.6. Isomorphism thus describes only a similarity of structures as wholes and of the position of corresponding elements within the structure, but says nothing about any other properties of the corresponding elements apart from their position in the structure. Such individual properties of the elements from which the structure is built are totally irrelevant for the question of whether the two structures are isomorphous; and isomorphism may not only exist between structures made of different materials but even between material and immaterial structures so long as there exist any common formal attributes of the relations which connect the elements.

2.7. In the application of the concept of isomorphism to psychological problems there has been a good deal of confusion with regard to the terms or structures which might be said to be isomorphous. There are three such different structures, any pair of which might be and has been represented as the terms between which isomorphism prevails. There are :

  1. The physical order of the external world, or of the physical stimuli, which for the present purpose we must assume to be known, although our knowledge of it is, of course, imperfect.

  2. The neural order of the fibres, and of the impulses proceeding in these fibres, which, though undoubtedly part of the complete physical order, is yet a part of it which is not directly known but can only be reconstructed.

  3. The mental or phenomenal order of sensations (and other mental qualities) directly known although our knowledge of it is largely only a 'knowing how' and not a 'knowing that',[1] and although we may never be able to bring out by analysis all the relations which determine that order.

2.8. Our problem is determined partly by the fact that the first and third of these orders are *not* isomorphous, i.e., that the physical order differs from the phenomenal order. Although the problem would also exist if these two orders were isomorphous (if that is conceivable), we might never, or at least not for a long time, have become aware of its character if it were not for the fact of the difference of these two orders. While they are in some measure similar, and while we owe it to this similarity that we can find our way about the physical world, they are, as we have seen, far from being identical.

[1]G. Ryle, 1945.

2.9. The isomorphism which we have suggested to exist refers to the relation between the second and the third of these orders, i.e., to the relation between the neural and the phenomenal order. If this is correct, and if the first and the third of these orders are not isomorphous, it also follows that the first and the second cannot be isomorphous. (That the second cannot be be strictly isomorphous with the first also follows from the fact that strictly speaking it is a part of the first).

2.10. Isomorphism between two structures or orders does not imply isomorphism between any properties their elements may possess apart from their place in the structure. This needs special emphasis as the term isomorphism has been used by Boring[1] to describe a correspondence between individual mental events (i.e., parts of our third order) and individual physical and neural events. He speaks of 'isomorphic transmission' of some constant structural feature from the stimulus through the impulse to the sensation, and in this sense the concept of isomorphism would indeed, as he points out, be merely a form of the naïve conception against which Johannes Müller's theory of the specific energy of nerves was directed. It is possible that in the vague use made of the concept by the gestalt school this meaning has been mixed up with the other one, but it need hardly be stressed that it has nothing to do with the sense in which the concept is employed here.

2.11. It should be pointed out at once, however, that our use of the term isomorphism, though useful for the purposes of exposition at this stage, will in the end also prove somewhat inappropriate. We are at present concerned with the relations of an inferred order, the terms of which are unknown (since they are left without attributes if we regard all mental attributes as determined by relations), with an order which might be established between the known neural elements. We shall, in fact, come to the conclusion that the two orders are not merely isomorphous but identical and that to postulate a separate set of terms for the mental order would be redundant. But at this step in the exposition we shall content ourselves to ask whether a topological equivalent of the mental order can be reconstructed from physical elements.

2.12. Another misunderstanding to which the use of the conception of a one-to-one correspondence in the discussion of iso-

[1]E. G. Boring 1935, p. 244, 1936, pp. 574—575, and 1942, pp. 83–90.

morphism could give rise, should at once be met. It is the old idea that individual stimuli and individual nervous impulses are invariably and uniquely related with particular individual sensory qualities. This cardinal error which, it will soon be seen, has been the main obstacle to the understanding of our problem, follows by no means from the conception of isomorphism as used here. On the contrary, if the action of an impulse depends on the position of the fibre that carries it, in the whole system of connected fibres, it would seem at once probable that its effects will depend on what other impulses are proceeding at the same time. Although at any given time (or within any given structure) any particular group of impulses occurring at the same time will have the same significance, there is no reason to expect that the effects of a single impulse will be the same whether it appears in company with one group or with another group of other impulses.

2.13.   This particular misunderstanding of the idea of a one-to-one correspondence between impulse and sensation has been persistently and successfully criticized by the members of the gestalt school[1] under the name of the 'constancy hypothesis'. Their experimental work has amply confirmed that such an invariable connexion between individual impulse and elementary sensations does not exist.

2.14.   Closely connected with this 'constancy hypothesis' is the conception of an 'invariable core of pure sensation' which is supposed to be in some manner originally attached to the nervous impulse and to continue to exist independently of all the modifications of, and additions to, this basic quality which may be effected by experience or acquired relations. Bertrand Russell, e.g., explicitly states with reference to this that 'the essence of sensation . . . is its independence of past experience.'[2]

2.15.   It has, of course, long been a common place in psychology that a large part of the experienced content of the sensory qualities is the result of interpretation based on experience. But these relational determinants of sensory qualities have invariably been represented as mere modifications of, or additions to, an original

---

[1] W. Köhler, 1913, p. 52; K. Koffka, 1935, pp. 85 ff.; and D. Katz 1944 for a clear distinction between this 'constancy hypothesis' and the 'constancy phenomenon', i.e. the fact that different stimuli and different combinations of stimuli can produce the same sensory qualities.

[2] B. Russell, 1921, p. 144. Cf., also *ibid* p. 139.

core of pure sensation.[1] It will be the central thesis of the theory to be outlined that it is not merely a part but the whole of sensory qualities which is in this sense an 'interpretation' based on the experience of the individual or the race. The conception of an original pure core of sensation which is merely modified by experience is an entirely unnecessary fiction, and the same processes which are known to modify and alter the qualitative attributes of sensations can also account for the initial differentiation.

2.16. With this contention we do not mean to assert that the 'learning' process which can account for the determination of the order of sensory qualities takes place entirely or predominantly in the course of the development of the individual. In this sense our contention does not take side in the dispute between the 'nativists' and the 'empiricists'. But this dispute seems usually to involve also the distinct question whether the order of sensory qualities can be understood as having been formed by the combined experience of the race and the individual, or whether it must be regarded as something unaccountably and unexplainably existing apart from the effects which the environment exercises on the development of the organism. In this second sense our thesis belongs to the 'empiricist' position (see 5.15).

2.17. It might indeed be said that the whole theory of the formation of sensory qualities to be developed in the following pages is no more than an extension and systematic development of the widely held view that every sensation contains elements of interpretation based on learning, an extension by which the *whole* of the sensory qualities is accounted for as such an interpretation. It will be contended that in the course of its phylogenetic and ontogenetic development the organism learns to build up a system of differentiations between stimuli in which each stimulus is given a definite place in an order, a place which represents the significance which the occurrence of that stimulus in different combinations with other stimuli has for the organism. We shall see later in what sense and to what extent this 'classification' (as we shall call it) of the stimuli by the organism can be said to 'reproduce' the 'objective' relations between those stimuli in the physical world.

2.18. It should, however, at once be noted, although a fuller discussion of this must be postponed to a later point,[2] that when

[1]For an account of these historical antecedents see Chapter VI.
[2]See Chapter VIII below.

we claim to provide an 'explanation' this will never mean more than an 'explanation of the principle' by which phenomena of the kind in question can be produced. By such an 'explanation of the principle' we shall provisionally understand an explanation which not only confines itself to showing 'that such and such actions lie within the range of known physical actions, or that known physical phenomena produce effects similar to them',[1] but also that, though we may be able to explain the general character of the processes at work, their operation may be so complicated in detail as to place their full description forever beyond the power of the human mind.

2.19. The reason for confining ourselves to such an 'explanation of the principle' is, therefore, not only that in the present state of psychology and neuro-physiology the main need seems to be for a hypothesis suggesting a possible way in which the phenomena in question may be produced, but also that there appear to exist reasons which should make for man a *full* explanation of his own processes of thought absolutely impossible, because this conception involves, as we hope to show, a contradiction.

2. THE ORDER OF SENSORY QUALITIES IN ITS STATIC AND ITS DYNAMIC ASPECTS

2.20. It is necessary now to examine a little more carefully than we have yet done the character of the various 'relations' existing between the sensory qualities. It would seem at first as if the fact which we have pointed out (1.56–1.61), that these 'relations' possess themselves different qualitative attributes, would constitute an absolute obstacle to any attempt to reproduce an equivalent or isomorphous physical system built up from the known physiological processes, since in the latter case the different elements can be ordered solely by the one relation of cause and effect. We have provisionally met this difficulty by pointing out that differences in quality can also be reduced to differences in the effects, but it clearly needs yet more explicit consideration.

2.21. This problem is closely connected with what may be called

---

[1]D. W. Thompson, 1942, p. 309. Cf., also E. G. Boring, 1946, p. 178 where he argues that 'it is enough for our purpose if we can produce the function *in kind*' and 'if we could *get the principle* of [the suggested synthetic professor of psychology] without actually producing him.' (Italics ours.)

the difference between the 'static' and the 'dynamic' aspect of the system of sensory qualities. We usually think of all the different sensory qualities as (at least potentially) existing at the same time, and it is this imagined simultaneous existence to which we refer when we speak of the 'static' aspect of the whole order. But as we have tried to show (1.38–1.55), all the questions which we can meaningfully ask about the differences between these qualities must necessarily refer to the different effects which in different combinations they will exercise on succeeding events: on how their appearance in a given situation affects our estimation of the other elements of the situation and so on. This is the system of qualities seen in its dynamic aspect. We shall later (2.44, 3.51, 5.42) see that the neural counterpart of the system of sensory qualities can similarly be regarded under the static aspect of an apparatus capable of performing the various discriminations, or dynamically by describing the various processes which it can perform.

2.22. Even when we imagine the system of sensory qualities as existing as a whole at a given moment, we do not mean that we ever have images of all the possible sensory qualities. What we mean when we think of that system as complete at any moment is that we could, as it were, run through it, proceeding from one quality to similar qualities, and by thus moving along all the possible dimensions, ultimately exhaust all possibilities. Even the 'static' system is thus in fact a sequence of images causally connected in complex ways.

2.23. The validity of the contention that all that can become a problem are the different effects which the different qualities produce, will depend on what in this connexion we include under 'effects'. If, with the strict behaviourists, we were to confine the term 'effects' to externally observable behaviour (overt action or other peripheral responses) the contention could certainly not be defended. There is no justification, however, for that exclusive concentration on overt action which, under the influence of behaviourism, has been the fashion in psychology during the last thirty years. Physiological research during the same period has rather made it clearer than ever that we cannot hope to account for observed behaviour without reconstructing the 'intervening processes in the brain.'[1]

2.24. It would indeed be absurd to recognize differences in the

[1]C. T. Morgan, 1943, p. 476.

responses only in so far as they manifest themselves in overt behaviour and to disregard our subjective knowledge of discrimination: not only because such an attempt could not be carried through consistently(1.84–1.88), but also because we know that the central nervous system provides an apparatus for just the kind of processes which, although they elude direct observation, can be shown to be necessary to bring about the observable results.[1] Any attempt to explain the distinction between sensory qualities in terms of peripheral responses was bound to fail, because there are no unique responses attached to particular stimuli. As we shall presently see, there is a process of multiple classification inserted between stimulus and response which makes it possible for the response to take account of the significance which the stimulus has in the context of other (external and internal) stimuli.

2.25. We shall have to show later (4.35—4.41) how this exclusive emphasis on peripheral responses is also misleading because, even in so far as peripheral responses contribute to the discrimination between stimuli, they can affect the further course of the mental processes only through the proprioceptive impulses (the 'feedback') by which they in turn are centrally recorded; even in these instances the decisive factors are therefore not the motor responses themselves, but the sensory impulses which they send back to the higher centres. We shall then see that it is also at least highly probable that, once a direct connexion has been established between the initial sensory impulse and the impulse recording the motor response evoked by it, the actual motor response becomes unnecessary for the continued functioning of this particular mechanism.

2.26. Once we include among the 'effects' of a stimulus all the intermediate links which may intervene between the stimulus causing a sensation and the overt response to it, the difficulty of defining sensory qualities in terms of their effects largely disappears. Whether we speak in terms of the physiological processes or in terms of the sensory qualities which they evoke, we shall find that the relevant differences between the individual events consist in the different immediate effects which they produce in different combinations. Each event or group of events will be distinguished from most others by the fact that it will evoke a particular set of other events. The ultimate overt response may thus be brought

[1]C. C. Pratt, 1939, p. 147.

about *via* a long series of intermediate links which in the neural process cannot be directly observed but can only be reconstructed from what we know of the mental counterparts of these processes and of those marginal overt responses to which the latter lead.

2.27. The apparent paradox that certain relations between non-mental events should turn them into mental events resolves itself as soon as we accept the definition of mind as a peculiar order. Any individual neural event may have physical properties which are similar or different from other such events if investigated in isolation. But, irrespective of the properties which those events will possess by themselves, they will possess others solely as a result of their position in the order of inter-connected neural events. As an isolated event, tested for its effects on all sorts of other such events, it will show one set of properties and therefore have to be assigned a particular place in the order or classification of such single events; as an element of the complete neural structure it may show quite different properties.

2.28. That an order of events is something different from the properties of the individual events, and that the same order of events can be formed from elements of a very different individual character, can be illustrated from a great number of different fields. The same pattern of movements may be performed by a swarm of fireflies, a flock of birds, a number of toy balloons or perhaps a flight of aeroplanes; the same machine, a bicycle or a cotton gin, a lathe, a telephone exchange or an adding machine, can be constructed from a large variety of materials and yet remains the same kind of machine within which elements of different individual properties will perform the same functions. So long as the elements, whatever other properties they may possess, are capable of acting upon each other in the manner determining the structure of the machine, their other properties are irrelevant for our understanding of the machine.[1]

2.29. In the same sense the peculiar properties of the elementary neural events which are the terms of the mental order[2] have nothing to do with that order itself. What we have called physical

[1] We are deliberately not using here the even greater number of examples of an order existing irrespective of the character of the elements of which it consists in which in any way mental factors are involved, such as, e.g., in the relation between a poem in its printed and in its spoken form, etc.

[2] E. G. Boring, 1933, p. 233.

properties of those events are those properties which will appear if they are placed in a variety of experimental relations to different other kinds of events. The mental properties are those which they possess only as a part of the particular structure and which may be largely independent of the former. It is at least conceivable that the particular kind of order which we call mind might be built up from any one of several kind of different elements—electrical, chemical, or what not; all that is required is that by the simple relationship of being able to evoke each other in a certain order they correspond to the structure which we call mind.

2.30. That a particular order of events or objects is something different from all the individual events taken separately is the significant fact behind the endless and unprofitable talk about 'the whole being greater than the mere sum of its parts'. Of course an order does not arise from the parts being thrown together in a heap, and one arrangement of a given set of parts may constitute something different from another arrangement of the same set of parts. An order involves elements *plus* certain relations between them, and the same order or structure may be formed by any elements capable of entering into the same relations to each other. The capacity of entering into such a relation is, of course, a property of the elements as much as any of those other properties which are irrelevant so far as the particular order is concerned. A particular order can exist as little without elements possessing that capacity, as the elements without the order in which they are related to each other would possess the particular significance which they have in that order. But it is only when we understand how the elements are related to each other that the talk about the whole being more than the parts becomes more than an empty phrase. All that theoretical biology has in this respect to say on the significance of structural properties as distinct from the properties of the elements, and about the significance of 'organization'[1], is directly applicable to our problem.

2.31. The question which thus arises for us is how it is possible to construct from the known elements of the neural system a structure which would be capable of performing such discrimination in its responses to stimuli as we know our mind in fact to perform.

[1]J. H. Woodger, 1929, p. 291 and *passim*; L. von Bertalanffy, 1942 and 1949.

3. THE PRINCIPLE OF CLASSIFICATION

2.32. The phenomena with which we are here concerned are commonly discussed in psychology under the heading of 'discrimination'. This term is somewhat misleading because it suggests a sort of 'recognition' of physical differences between the events which it discriminates, while we are concerned with a process which *creates* the distinctions in question. The same is true of most of the other available words which might be used, such as 'to sort out', 'to differentiate', or 'to classify'. The only appropriate term which is tolerably free from misleading connotations would appear to be 'grouping'.[1]

2.33. For the purposes of the following discussion it will nevertheless be convenient to adopt the term 'to classify' with its corresponding nouns 'classes' and 'classification' in a special technical meaning. The next few paragraphs will serve solely to make precise the exact meaning in which we propose to use this term. We shall at first consider extremely simple processes of classification which will have little resemblance to the more complex kinds which are relevant to our main task. Our present purpose will be more to make clear what the principle of classification as such involves, than to show how it operates in the nervous system.

2.34. By 'classification' we shall mean a process in which on each occasion on which a certain recurring event happens it produces the same specific effect, and where the effects produced by any one kind of such events may be either the same or different from those which any other kind of event produces in a similar manner. All the different events which whenever they occur produce the same effect will be said to be events of the same class, and the fact that every one of them produces the same effect will be the *sole* criterion which makes them members of the same class.

2.35. We may conceive of a machine constructed for the purpose of performing simple processes of classification of this kind. We can, for instance, imagine a machine which 'sorts out' balls of various size which are placed into it by distributing them between different receptacles. We will assume that no two balls have the same size so that size is merely a means of identifying the individual balls. Indeed we shall even assume that no two balls have any

---

[1]'Grouping' was used somewhat in this same sense by G. H. Lewes, 1880, Problem III, Chapter 3, §§ 33 and 34, and more recently by J. Piaget, 1947.

property in common which they do not share with every other ball in the set, so that there are not 'objective' similarities peculiar to the different members of any subgroup or class of these balls; any grouping of different balls by the machine which places them into the same receptacle will create a class which is based exclusively on the action of the machine and not on any similarity which those balls possess apart from the action of the machine.[1]

2.36. We may find, for instance, that the machine will always place the balls with a diameter of 16, 18, 28, 31, 32, and 40 mm in a receptacle marked A, the balls with a diameter of 17, 22, 30, and 35 in a receptacle marked B, and so forth. The balls placed by the machine into the same receptacle will then be said to belong to the same class, and the balls placed by it into different receptacles to belong to so many different classes. The fact that a ball is placed by the machine into a particular receptacle thus forms the sole criterion for assigning it to a particular class.

2.37. Another kind of machine performing this simplest kind of classification might be conceived as in a similar fashion sorting out individual signals arriving through any one of a large number of wires or tubes. We shall regard here any signal arriving through one particular wire or tube as the same recurring event which will always lead to the same action of the machine. The machine would respond similarly also to signals arriving through some different tubes or wires, and any such group to which the machine responded in the same manner would be regarded as events of the same class. Such a machine would act like a simplified telephone exchange in which each of a number of incoming wires was permanently connected with, say a particular bell, so that any signal coming in on any one of these wires would ring that bell. All the wires connected with any one bell would then carry signals belonging to the same class.

2.38. An actual instance of a machine of this kind is provided by certain statistical machines for sorting cards on which punched holes represent statistical data. If we regard the appearance of any card with the same data punched on it as the recurrence of the same event, and assume that the machine is so arranged that various groups of different data are placed into the same receptacle,

[1] J. Piaget, 1947, p. 45: 'Un concept de classe n'est psycholiquement que l'expression de l'identité de réaction du sujet vis-a-vis des objets qu'il reunit en une classe.'

we should have a machine which performs a classification in the sense in which we use this term.

4. MULTIPLE CLASSIFICATION

2.39. In the kind of simple classification which we have just considered, any one of the individual recurrent events is always grouped with the same group of other events and with them only. But the same principle can effect what may be called multiple classification: at any moment a given event may be treated as a member of more than one class, each of these classes containing also different other events; and a given event may also on different occasions be assigned to different classes according to the accompany events with which it occurs. The classification may thus be 'multiple' in more than one respect. Not only may each individual event belong to more than one class, but it may also contribute to produce different responses of the machine if and only if it occurs in combination with certain other events. Different groups consisting of different individual events may in this manner evoke the same response and the machine would then classify not only individual events but also groups consisting of a number of (simultaneous or successive) events. In this latter case the groups (or sequences) of individual events would as groups constitute the elements of the different classes.

2.40. The first kind of multiple classification could be performed, for instance, by a machine similar to the first we have imagined if, instead of placing the balls into different receptacles, it were to show different signs, say lights of different colours, every time a ball is placed into it. A ball to which the machine responded by showing a red and a green light would then belong to two classes of balls, that of all balls evoking a red light and that of all balls evoking a green light. Or, in the case of the second kind of machine described before, which performs the classification by establishing connexions with different bells, each incoming signal might be passed on to more than one bell and belong accordingly to a corresponding number of different classes.

2.41. The second type of multiple classification would be represented by a machine whose responses depended not only on the individual events to be classified but also on the combinations in which they occurred. The classification of the groups of events by such a machine might be either additional to the classification of the individual events, or occur in the place of it, so that the individual event which, if it occurred in isolation, evoked say a green light, would not do so but contribute to produce a blue light if it occurred at the same time with, or within a short interval of certain other events.

2.42. We shall later (3.52–3.57) have to consider yet a third type of multiple classification: namely one in which successive acts of classification follow upon each other in relays, or on different 'levels'; in this type the distinct responses which effect the grouping at a first level become in turn subject to a further classification (which also may be multiple in both the former senses). This is probably the most important characteristic of the particular kind of classificatory mechanism which the nervous system represents; but while we are merely concerned to bring out certain general principles, we shall disregard this aspect until the next chapter.

2.43. In the system of classification in which we shall be interested the different individual events will be the recurrent impulses arriving through afferent fibres at the various centres of the nervous system. For the purposes of this discussion we shall have to assume that these individual impulses possess no significant individual properties which distinguish them from one another. They must be regarded initially as what the logician describes as an 'uninterpreted set of marks'. Our task will be to show how the kind of mechanism which the central nervous system provides may arrange this set of undifferentiated events in an order which possesses the same formal structure as the order of sensory qualities.[1]

2.44. Throughout the discussion of that neural apparatus of classification it will be important to keep in mind the distinction between the structural and the functional (or the static and the dynamic) aspect of that mechanism (2.20–2.31). The elements of the (anatomical) structure will be the different fibres; the element of the (physiological) process will be the impulses conducted by these fibres. It will be the impulses which (as individuals or groups) will be the object of the classificatory process.

2.45. Our task will thus be to show how these undifferentiated individual impulses or groups of impulses may obtain such a

[1]For a somewhat similar statement of the problems of the order of sensory qualities see R. Carnap, 1928.

position in a system of relations to each other that in their functional significance they will resemble one another or differ from another in a manner which corresponds strictly to the relations between the sensory qualities which are evoked by them.

5. THE CENTRAL THESIS

2.46. We shall maintain that a classification of the sensory impulses which produces an order strictly analogous to the order of sensory qualities can be effected by a system of connexions through which the impulses can be transmitted from fibre to fibre ; and that such a system of connexions which is structurally equivalent to the order of sensory qualities will be built up if, in the course of the development of the species or the individual, connexions are established between fibres in which impulses occur at the same time.

2.47. That such connexions through which impulses are transmitted are created as a result of the simultaneous occurrence of sensory impulses is an almost universally accepted hypothesis which seems indeed indispensable if we are to account for such well-established facts as conditioned reflexes,[1] even though we do not yet know exactly how they are established or maintained. For the purposes of our argument it is irrelevant whether the establishment of such connexions involves, as used to be generally assumed, a change in the anatomical structure of the central nervous system (such as the 'formation of new paths'), or whether, as some more recent investigations suggest, they are based on physiological or functional changes, such as the setting up of some continuous circular flow of impulses in certain pre-existing channels.[2]

2.48. The transmission of impulses from neuron to neuron within the central nervous system, which is thus conceived as constituting the apparatus of classification, may either take place between different neurons carrying primary impulses, or between such neurons and other ('internuncial') neurons which are not directly connected with receptor organs. In the former instance the same

[1]More recently the occurrence of such connexions between sensory impulses has also been established by psychological experiments by W. J. Brogden 1939, 1942, 1947 and 1950.
[2]For an account of these newer views see E. R. Hilgard and D. G. Marquis, 1940, p. 330.

event, an impulse in an afferent neuron, may occur either as the primary object of classification or as a 'symbol' classifying some other primary impulse. But since, as we shall see, all impulses, whether primary or secondary in this sense, are likely to be subject to further acts of classification, and therefore to appear both as instruments and as objects of classification, this merely complicates the picture but does not alter the general character of the process.

2.49. The point on which the theory of the determination of mental qualities which will be more fully developed in the next chapter differs from the position taken by practically all current psychological theories[1] is thus the contention that the sensory (or other mental) qualities are not in some manner originally attached to, or an original attribute of, the individual physiological impulses, but that the whole of these qualities is determined by the system of connexions by which the impulses can be transmitted from neuron to neuron; that it is thus the position of the individual impulse or group of impulses in the whole system of such connexions which gives it its distinctive quality; that this system of connexions is acquired in the course of the development of the species and the individual by a kind of 'experience' or 'learning'; and that it reproduces therefore at every stage of its development certain relationships existing in the physical environment between the stimuli evoking the impulses. (We shall see in Chapter IV that this 'physical environment' within which the central nervous system operates includes the *milieu intérieur*, i.e., the organism itself in so far as it acts independently of the higher nervous centres; and in Chapter V how this 'experience' differs from experience in the ordinary meaning of the word.)

2.50. This central contention may also be expressed more briefly by saying that 'we do not first have sensations which are then preserved by memory, but it is as a result of physiological memory that the physiological impulses are converted into sensations. The connexions between the physiological elements are thus the primary phenomenon which creates the mental phenomena.'[2]

---

[1]The closest approximation to the theory developed here seems to have been reached by D. O. Hebb, 1949, a work which came to my knowledge only after the present book was completed in all essentials.

[2]The quotation is a translation from the early German draft of the present work (1920) referred to in the Preface.

2.51. Although suggestions of a theory of mental phenomena on these lines are implicit in much of the current discussion of those problems by physiological psychologists, the consequences of a systematic application of this basic idea appear never to have been worked out consistently. What follows is little more than an attempt to elaborate the main implications of this thesis. It will be seen that its consistent development leads to rather important conclusions and assists in the clearing up of several old puzzles.

CHAPTER III

THE NERVOUS SYSTEM AS AN
INSTRUMENT OF CLASSIFICATION

1. AN INVENTORY OF THE PHYSIOLOGICAL DATA

3.1. Before we can attempt to state in greater detail the theory sketched in the preceding chapter, it will be necessary to take stock of the essential anatomical and physiological facts which we shall have to use as bricks from which to construct an apparatus of the kind we are seeking. For our purpose it will not be necessary to concern ourselves with the structure and the functioning of the central nervous system in any great detail. It will suffice if we briefly note certain general characteristics of its parts and of the processes taking place in them. The simplifications which we shall employ must be justified by the fact that our aim is not so much to elaborate a theory which is correct in every detail, as to show how any theory of this kind can account for the mental events with which we are concerned.

3.2. According to an almost universally held view the nervous system is built up, like the rest of the organism, from a large number of separate cells. These cells, called neurons, consist of a cell body and two kinds of attaching processes, the axon and the dendrites. Although some doubt has recently been expressed concerning this 'neuron theory', and the alternative theory of an essential continuity of the system of nervous fibres has been put forward,[1] we shall state the main facts in terms of the predominant view, since confirmation of the alternative theory would not significantly affect the conclusions at which we arrive from the former. The main facts which we shall have to take into account may then be stated as follows:

[1] For a brief summary of the recent German work on the alleged 'continuity of the nervous system' see W. Bargmann, 1947, and for a criticism, N. A. Hillarp, 1947.

3.3. The cerebral cortex is the highest and most complex of several 'bridges' which connect the afferent fibres conducting impulses from the peripheral receptors, and the efferent fibres conducting impulses to the motor organs. We must thus conceive of the central nervous system (and probably also of the cortex itself) as a hierarchy consisting of many superimposed levels of connexions, all of which may be concerned in the transmission of impulses from the afferent (sensory) to the efferent (motor) fibres. This conception of a hierarchy of centres or levels does, of course, not imply that these levels can always be sharply separated, either structurally or functionally, or that they are superimposed upon each other in a simple linear order.

3.4. The number of separate nerve cells within these centres by far exceeds the number of afferent fibres conducting impulses to them and of the efferent fibres conducting impulses from them. The cerebral cortex alone has been estimated to contain about ten thousand million separate cells while the number of afferent and efferent fibres is of the order of magnitude of a few millions only. The number of distinct afferent fibres reaching the cortex is also considerably lower than the number of distinct sensory receptors which are the source of the impulses reaching the brain through these fibres.

3.5. While the peripheral receptor organs in which the impulses are set up by stimuli are in general sensitive only to a limited range of stimuli, the impulses themselves which are conducted to the nervous centres are of uniform character and do not differ from each other in quality. There is no known correspondence between any attributes of the individual impulse and either the attributes of the stimulus which caused it or the attributes of the sensory quality which it evokes (1.31–1.37).

3.6. The impulse or state of excitation conducted by any nervous fibre is not a continuous flow but rather a succession of shocks following each other at very short intervals and usually described as a 'train' (or incorrectly as a 'volley') of impulses.

3.7. Each fibre will normally conduct impulses only in one direction, although it seems probable that the fibre itself is capable of transmitting impulses in either direction and that it is its position with respect to the body of the cell, and the position of the whole neuron in the chain of neurons, which determines in which direction the impulses will normally travel through a fibre.

3.8. The impulses conducted by the nerve fibres obey the 'all-or-nothing principle' which states that any given fibre may only either transmit or not transmit a given impulse, but that, if it does transmit it, the impulse will always be of the same strength. This means that we have throughout to deal with a kind of 'trigger phenomenon' where what is loosely called a 'transmission' of impulses does not really mean a conduction of energy but rather that one impulse releases energy stored up in the next cell in the chain.

3.9. The 'strength' of the impulse, which shows itself in its capacity to cause excitation in other neurons, however, will differ not only between different fibres but also between different segments and branches of the same fibre roughly in proportion to their thickness. But while the impulse conducted by a given fibre cannot vary in strength, it may vary in duration (or rather in the number of successive shocks of which the train of impulses is made up), and this variation in duration will in some respects operate similarly to a variation in strength (see 3.13 below).

3.10. In addition to the impulses transmitting excitation some nerve fibres appear to conduct another kind of impulses which quell or inhibit excitation.

3.11. At certain points called 'synapses' nervous impulses are transmitted from one neuron to another. Any theory that is to account for the known action of the central nervous system must assume that these 'synapses' are not permanent or invariable features of the nervous system but can be created and modified in the course of its operation, probably as a result of the simultaneous occurrence of impulses in two or more adjoining neurons. As has already been pointed out (2.47), we possess practically no knowledge about the nature of these synapses or the mechanism by which they are created. It is not even clear whether we ought to conceive of the creation of a new synapse as a change in the anatomical structure, which is the interpretation commonly given to the 'formation of a new path',[1] or whether it is brought about by a functional change, such as the establishment of the kind of permanent circular flow of impulses mentioned before. In so far as

---

[1] Cf. E. D. Adrian, 1947, p. 92: 'The notion that memories might be related to structural changes of this kind has often been rejected on the ground that no one has been able to detect them with the microscope, but the chance of doing so would be so remote that the objection need not be taken very seriously.'

connexions of this kind must be assumed to transmit not excitation but inhibition, there does not appear to exist even a plausible
hypothesis about the conditions under which such new connexions
would be established, comparable to the rôle attributed to the
simultaneity of the impulses for the formation of connexions between excitatory impulses.

3.12. The assumption that connexions or synapses between
neurons are created as the result of the simultaneous excitation of
these neurons implies the further assumption that these connexions
will be two-way connexions, i.e., that, if an impulse in a given
neuron is regularly transmitted to a certain other neuron, an
impulse in this second neuron will also be regularly transmitted
to the first. This assumption is independent of the question
whether the transmission in the two opposite directions is effected
by the same channel or whether separate channels capable of
transmitting impulses in opposite directions are created by the
same circumstances.

3.13. The operation of the 'all-or-nothing principle' is partly
modified by the phenomenon of 'summation' which appears to
operate in two ways, spatially and temporally: either impulses
arriving simultaneously at a given cell through different fibres,
although each of them individually may be too weak to cause
excitation of that cell, may yet together produce that result; or the
succession of shocks contained in a train of impulses in a single
fibre may build up sufficient strength to cause excitation to the
cell to which they are conducted, although a single shock or a few
shocks would not have been sufficient to do so.

3.14. It seems that in many instances the stimulation of more
than one individual receptor organ and sometimes perhaps the
stimulation of receptor organs of several different kinds, and
consequently the arrival of impulses through a number of different
afferent fibres, is required in order that a sensation of a particular
quality should be produced.[1]

2. SIMPLIFYING ASSUMPTIONS ON WHICH THE OPERATION OF THE PRINCIPLE WILL BE DISCUSSED

3.15. In the preceding enumeration of some of the main features
of the functioning of the central nervous system certain facts have

[1] C. T. Morgan, 1943, pp. 297–8.

been deliberately left out which are not required for the very simplified account of its functioning as an instrument of classification which will be attempted here. In particular, we have left out much that would be important if we were to attempt to sketch the temporal pattern of the order of impulses. But although there can be no doubt that this time structure is very important, any attempt to describe it would have to make use of a great deal more of physiological detail than would be compatible with a clear presentation of the outline, or would be justified by the present state of our knowledge of these matters.[1]

3.16. Even when we leave out this problem of the temporal order of the neural events, the possibilities of classification of impulses which the structure of the neural system provides are of such a manifold character that, in order to obtain a clear picture of how the principle operates, it will be advisable to approach the actual situation by gradual steps. We shall therefore at first employ a number of simplifying assumptions which will later be gradually dropped. The simple models which we shall discuss in the present chapter serve merely to bring out certain salient features of the complex process of classification.

3.17. The first simplifying assumption of this kind which we shall employ provisionally is that we shall consider how a single afferent impulse arriving at the higher centres may here be classified or be discriminated from other similar impulses. This is, of course, a very artificial case, since it is most unlikely that at any moment only one such impulse will arrive, and even doubtful whether, if this ever happened, such an isolated impulse would give rise to a sensation.

3.18. The second simplifying assumption we shall adopt for the present is perhaps even more drastic and unrealistic. We shall concentrate entirely on the order created by connexions formed between sensory neurons and for the time being entirely neglect the connexions established between sensory and motor neurons. The whole problem of the relation between sensation and motor action or behaviour will be taken up only in the next chapter.

3.19. Closely connected with this second simplification is a third which we also shall adopt for the time being, namely the disregard of the hierarchal structure of the central nervous system. We shall, in other words, begin by considering how connexions between

[1]See, however, now D. O. Hebb, 1949.

sensory neurons might create an order if they were all formed in a single centre or on one and the same level.

3.20. The two last-mentioned simplifications mean, of course, that as a first approximation we shall neglect two facts which are of crucial and decisive importance for the actual functioning of the nervous system. It has rightly become a commonplace in neurophysiology that we must not think in terms of separate sensory and motor mechanisms but rather in terms of a single sensorimotor system. If, nevertheless, at first we treat in isolation that part of the sensory order which might be produced by connexions between the sensory impulses only, and postpone to the next chapter the questions of the interaction between sensory and motor impulses, this is in deliberate contrast to current practice. Our procedure is based on the belief that in recent times the direct connexions between sensory and motor impulses have been rather overstressed at the expense of an adequate recognition of the order which may be determined by connexions within the sensory sphere only.

3.21. When in the course of this chapter we speak of the 'effects' of particular sensory impulses we shall therefore refer to their effects on other central processes. These effects may consist in the evocation of other impulses either in neurons which can also be excited by primary impulses, or of impulses in 'internuncial' neurons in which an impulse acts, as it were, merely as a symbol or sign for a class of afferent impulses.

3.22. We shall also, for the purposes of the present discussion, continue to disregard one of the main difficulties which a fuller examination of our problem would have to face: the distinction between the phylogenetic and the ontogenetic aspects of the process of the formation of the order of sensory qualities. As we have already mentioned (2.49), it is probable that some of the connexions formed in the development of the species become embedded in the structure of the central nervous system while others will be formed during the life of the individual. For the purposes of the present schematic sketch we shall neglect this distinction and proceed as if the formation of the system of connexions commenced in an individual organism endowed with an apparatus capable of forming such connexions but in which at the outset no such connexions existed.

3.23. Another important question which for lack of sufficient knowledge we must leave undecided, is whether the connexions formed between neurons which simultaneously receive afferent impulses will be direct connexions between these neurons or whether we ought to conceive of them as mediated by other cells which are not directly linked with receptor organs but serve merely as connecting links between other neurons. Such third-cell connexions certainly occur, and from the proportion between the total number of neurons in the cortex and the much smaller number of afferent and efferent fibres (3.4) it would appear that the greater part of the neurons forming the cerebral cortex can have no direct connexions with receptor or effector organs and are likely to perform some such mediating function.

3.24. Finally, it should be remembered throughout the following discussion that when we speak of connexions this will include what we may call 'potential' as well as effective connexions, i.e., connexions which transmit impulses which by themselves would not be strong enough to cause excitation of the neurons to which they are conducted, unless they are reinforced (through summation) by other impulses arriving there more or less at the same time, as well as connexions carrying impulses sufficiently strong by themselves to transmit excitation.

3. ELEMENTARY FORMS OF CLASSIFICATION

3.25. If we now turn to consider the significance of the fact that the different sensory neurons in the cortex will have acquired various sets of connexions with other neurons, it will at once be evident that if each of two or more neurons should be connected with exactly the same other neurons, so that an impulse occurring in any one of the former will be transmitted to the same group of other neurons, the effects of an impulse in any one of the former will be the same. Their position in the whole structure of connexions would be identical and their functional significance would be the same. (Cf., 2.5).

3.26. With this extreme instance of complete identity of all connexions possessed by a number of neurons we may at once contrast the opposite instance where a number of neurons possess no common connexions with the same other neurons. Between these two limiting cases there may exist any number of intermediate positions: groups of neurons which have a larger or smaller part of their connexions in common. We can thus speak of greater or smaller degrees of similarity of the position of the different neurons in the whole system of such connexions.

3.27. This similarity of the positions of the individual neurons in the whole system of connexions can vary not only in degree but also in kind. Of three neurons, a, b, and c, possessing the same number of connexions with other neurons, a may have the same number of connexions in common with b as it has with c, which would mean (at least if all these connexions were also of the same strength) that the similarities between the positions of a and b and between the positions of a and c were of the same degree. Yet these similarities might be of different kinds, because some or all of the connexions which a had in common with b might be different from those which a had in common with c. This means, of course, that although the position of a in the whole system of connexions would be similar to that of b and to that of c, there might be much less similarity or no similarity at all between the positions of b and c. This merely expresses the fact that the relation of similarity is non-transitive (1.46).

3.28. A very high degree of similarity in the position of the different neurons in the system of connexions is likely to exist wherever the neurons are served by receptors sensitive to stimuli which always or almost always occur together. This is most likely where these receptors are not only sensitive to the same kind of physical stimuli but also situated in close proximity.

3.29. If we can show how all the afferent impulses which give rise to sensations of the same quality are likely to be transmitted to the same group of further neurons, and by this fact will be distinguished from impulses producing different sensory qualities, we shall have provided an answer to our problem in the simplest form in which it occurs: we shall have explained the equivalence of the impulses occurring in different fibres. There are several obvious reasons which lead us to expect that such a classification of certain impulses as equivalent in all or some respects will be brought about as a result of the relative frequency with which different impulses occur together.

3.30. In the first instance, it is on the whole more likely that receptor organs sensitive to physically similar stimuli will be excited at the same time, and it is therefore to be expected that especially close connexions will be formed between the central neurons to which the corresponding impulses are transmitted. Where the physical stimuli can vary continuously in one or more dimensions, as in the case of light or sound, mixtures or bands of various frequencies of light or sound waves usually occur together and those which are more closely similar in a physical sense probably also occur more frequently together. It is thus to be expected that, e.g., not only all impulses set up by light waves (or by sound waves) will acquire some common connexions but also that there will be more such common connexions according as these stimuli are more or less closely akin physically.

3.31. These connexions are likely to be closest where the receptors are situated near to each other, but we shall also expect all the receptors of a given organism which are sensitive to the same kind of physical stimuli to be frequently excited at the same time, so that a fairly dense net of connexions will be formed between the corresponding central neurons. Similarly we shall expect fairly close connexions to be formed between the neurons served by neighbouring receptors which are sensitive to stimuli which occur frequently together because they emanate from the same physical objects, such as pressure and temperature, certain chemical agents acting simultaneously on mouth and nose, etc., etc.

3.32. Secondly, any particular kind of stimulus will usually occur more frequently in the company of some other stimuli than in that of others, and the connexions between the central neurons corresponding to physically different stimuli will thus come to reflect the relative frequency in which these different stimuli occur together. What has been said before about the specially close connexions between impulses caused by physical stimuli of the same kind will also apply to impulses caused by stimuli which, although they are not, like all light waves, physically closely similar, at least, like movement and sound, usually occur together.

3.33. Thirdly, in many instances it is likely that certain kinds of stimuli will usually act together on the organism when the organism itself is in a particular state of balance or of activity, either because the stimulus regularly occurs under conditions producing that state, or because it occurs periodically so as to coincide with some rhythm of the body. The impulses which register such external stimuli will then become connected with impulses received from the proprioceptors which register the different states of the organism itself.

3.34. The result of all this will be that a system of connexions will be formed which will record the relative frequency with which in the history of the organism the different groups of internal and external stimuli have acted together. Each individual impulse or group of impulses will on its occurrence evoke other impulses which correspond to the other stimuli which in the past have usually accompanied its occurrence. We shall call this bundle of secondary impulses which each primary impulse will set up through these acquired connexions the *following* of the primary impulse. It will be the total or partial identity of this following of the primary impulse which makes them members of the same class.[1]

4 · COMPLEX FORMS OF CLASSIFICATION

3.35. Even as a result of the comparatively simple processes discussed in the last section each impulse would become the member not merely of one class but of as many distinct classes as will correspond, not only to the number of other impulses which constitute its following, but in addition also to the number of possible combinations (pairs, triples, quadruples, etc.) of such other impulses; it might have any such part of its following in common with different groups of other impulses and therefore form a distinct class with them. We obtain thus already a somewhat complex form of 'multiple' classification in the first of the senses distinguished before (2.39–2.40).

3.36. Attention should be directed already at this stage to a circumstance which will have to be further considered at a later point (3.52ff.), namely, the fact that the kind of classificatory processes which we are now considering differ from those performed by the machines discussed earlier in one important respect. In the instances which we are now considering, the classificatory responses are not different in kind from, but are events of the same sort as, those which are the object of classification. In consequence,

---

[1] We cannot, without going more deeply into the physiological problems involved than seems expedient, examine the question whether the group of connected impulses which thus form the following of any particular primary impulses may not come to form relatively stable aggregations in the sense that, by the individual impulses mutually evoking each other, they may maintain themselves for some time beyond the duration of the stimulus. Such a conception appears to underlie the construction of a 'cell assembly' used by D. O. Hebb, 1949.

it is possible that one and the same event may appear both as an object of classification and as an act of classification. The impulse produced by a peripheral stimulus is 'classified' by evoking other impulses which might also be produced by peripheral stimuli. We shall see that it is a consequence of this relationship between the classifying and the classified impulses that a process of classification can produce 'models' of extremely complex relationships between stimuli, and indeed can reproduce the order of any conceivable structure.

3.37. In accordance with the distinction we have drawn between 'effective' and 'potential' connexions between neurons (3.24) we shall also have to distinguish between that part of the following of an impulse which will always occur whenever that impulse occurs, and that part which is merely 'potential' and which will appear only if the tendency towards an excitation of the neurons constituting the 'potential' following is supported by other impulses operating towards the same effect.

3.38. In the extreme case where the following acquired by any one neuron of a given class is completely identical with that of the other members of the class, their individual position in the whole system of connexions and therefore their functional significance would also be identical. This result is possible but not likely to occur frequently. When I was first working on these problems I thought I had found an instance of such undistinguishable sensations caused by stimuli operating on different receptors in the case of pressure on teeth standing opposite each other; I am no longer in a position to verify this. Such pressure seemed to me indistinguishable so far as immediate experience was concerned, and I was able to decide which tooth was concerned only by calling in further sensory experience, such as touching the teeth individually with my fingers. This case would, of course, satisfy the condition that the two stimuli almost always occur together.

3.39. There are undoubtedly other instances where stimuli, although they set up impulses in distinct fibres, remain indistinguishable. As a rule, however, we find that even very similar sensations caused by the stimulation of different receptors differ from each other, if in no other way, at least by an awareness of the different points at which they occur, or by what used to be called their 'local sign'. If we are to account for these differences between the effects of impulses which produce sensation otherwise of the same quality, we shall, in addition to the common following which accounts for their similar quality, have to find differences which account for their assignment to different points in space.

3.40. If we examine this problem at first in connexion with vision we enter a field where the kind of explanation which we are attempting to apply generally was first used to account for a special problem: ever since Bishop Berkeley the connexions between the impulses registering the visual stimuli on the retina and the kinesthetic impulses recording the tension of the muscles used for focusing the eye have been employed for explaining the spatial order of sensations. A particular visual impulse may have acquired exactly the same connexions with other visual impulses and thus be classified as differing qualitatively in the same manner from all other visual stimuli, and yet it may differ from the former by being connected with a different set of kinesthetic impulses.

3.41. The use we shall make of this fact, however, will in two respects differ from that made of it in the Berkeley-Helmholtz-Mach theory of spatial vision. Firstly, the impulses registering the state of muscular tension will not be conceived as producing distinct sensations but will be considered merely as physiological events which are associated with, and evoked by, the visual stimuli, and which contribute to the peculiar effects which the latter are capable of producing.

3.42. Secondly, what we shall regard as connected with the visual impulses will not be the actual movements of the eye muscles but merely the sensory impulses which normally record such movements in the central nervous system but which may also occur, if they are associatively evoked by the visual impulses, without the eye movements actually occurring. It would, therefore, not be a valid objection against this interpretation if it were pointed out that the movements of the eye postulated by the traditional theory do not, in fact, take place.

3.43. The theory of spatial vision serves here merely as an example of the manner in which the spatial order of sensations can in general be accounted for. And even this in turn is significant for us mainly as an illustration of the even more general way in which most specific acts of sensation require particular postures or attitudes of the body in order that the characteristic quality of the sensation should be produced. This fact will have to be considered further in the next chapter (4.35–4.44).

3.44. At this point the artificial separation of the connexions existing between different sensory impulses from those between them and motor impulses becomes difficult to maintain. We have already been forced to take into account the motor responses usually accompanying the acts of perception—motor responses which are probably directed from some lower level of the central nervous system, and which in turn will give rise to kinesthetic and other proprioceptive impulses which will become part of the following of the initial sensory impulse. These matters, however, can only be considered further at a later stage.

3.45. The common spatial order,[1] which is part of the common order of all sensations, serves here merely as an instance of the great variety of relations between different sensory impulses which will help to build up that order. The bundles of connexions or the following which any one of several sensory neurons may acquire can differ from those of others in an almost infinite variety of ways, ranging from complete identity of their following to complete absence of any common connexions; and every difference in the connexions which the individual neurons possess will have its peculiar functional significance.

3.46. To the possibilities of functional differentiation of the different impulses by connexions transmitting excitation, we have to add the effects of the second kind of connexions mentioned earlier, namely those transmitting inhibition (3.10). In whatever manner such inhibitory connexions may be acquired in the first instance and later transferred, their existence extends the range of possible differences in the position which any one impulse may occupy in the whole system of connexions: it adds the possibility of different impulses having effects which are directly opposed to each other. In such instances the evocation of certain other impulses normally following upon the occurrence of a given impulse would be prevented if still other impulses occurred at the same time. The range of functional differentiation of the impulses which may be determined by the differences in their following is thus

[1]It is significant for our purposes that this common spatial order extends only as far as the same physical events give rise to the stimulation of different senses and that, e.g., as William James has pointed out, our perception of size within the cavity of the mouth, which in the nature of the case is not co-ordinated with visual stimuli, does not fully form part of the same spatial order as our visual experiences.

extended, from equality through various degrees of similarity and difference, to contrariness and complete opposition.

3.47. The significance of the differences in the following which different neurons will have acquired will show itself in the differences between the effects which the impulses occurring in them will produce in different circumstances. The manner in which any newly arriving impulse will modify the existing excitatory state of the whole nervous system, and in which it will combine its effects with those of all other simultaneously arriving impulses, will depend on the different followings of all those impulses. The further course through which any one bundle of impulse-chains will run will be determined by the following of each successive impulse and by the manner in which the impulses of this following will combine with (i.e., reinforce or inhibit) other impulses proceeding at the same time (*see* 5.53).

3.48. In the complex interplay of many chains of impulses proceeding at the same time, the identity of the greater part of the followings which two or more neurons possess will bring it about that the occurrence of any one of them will in most situations produce the same or similar results, and that their simultaneous occurrence will tend to reinforce those parts of their following which they have in common.

3.49. Such a system, in which each of a set of events is connected with many others in such a way that the occurrence of any one or of any group of them causes (or contributes to bring about) the occurrence of certain others, evidently performs classifications in the sense in which we have defined this term. All the impulses or groups of impulses which evoke the same other impulses will belong to the same 'class' because they have this particular effect in common. Individual impulses or groups of impulses will of course almost always belong to a great many different classes, that is, multiple classification in the first of the different senses we have distinguished (2.40) will be the rule.

3.50. Since the different individual impulses will become members of a class through the fact that each of them evokes the same other impulses, it seems permissible to say that the latter *represent* the common attribute of the members of the class—though it would be more correct to say that they *constitute* that attribute. The classification is effected by the evocation of certain other impulses, and the latter serve, as it were, as the 'signs' or 'symbols' representing the class; the expression 'representative processes in the brain' which has been much used in recent physiological psychology,[1] can therefore appropriately be applied to them.

3.51. It has been suggested above (2.20–2.31, 2.44) that the mechanism which we are considering can be conceived either 'statically', as an apparatus capable of performing classifications, or 'dynamically', as a process of classification. In the preceding discussion we have sometimes spoken in terms of the former, e.g., when we spoke of connexions between neurons through which impulses are transmitted, and sometimes in terms of the latter, when we spoke of the impulses evoking each other. These two aspects of the same phenomenon correspond to the two aspects of the system of sensory qualities which we discussed then. It should now be clear that it is the dynamic aspect which is the really relevant one and that the static view is merely a method which is sometimes convenient to use for describing the potential operations of the system.

5. THE CLASSIFICATION OF THE RELATIONS BETWEEN CLASSES

3.52. There is no reason why such connexions as we have been considering should be formed only between primary sensory impulses, i.e. between impulses arriving through afferent fibres at the higher centres; they can evidently be formed in a similar manner between the further impulses which are evoked by the former and which represent classes of them. Any impulse which occurs as part of the following of one or more other impulses will, on each occasion when it thus occurs, acquire or strengthen connexions with other impulses which form part of the same following. Connexions of this kind will therefore also be formed between impulses which as primary impulses rarely if ever occur at the same time, but which on different occasions have become connected with the same third impulse, in the following of which they have in consequence become included.

3.53. This acquisition of connexions between impulses in consequence of their simultaneous occurrence in a secondary or derived character is specially important in so far as those neurons in the cerebral cortex are concerned which are not directly served by

[1] *See* C. T. Morgan, 1943, pp. 467, 476.

sensory receptors but which appear to act solely as intermediaries between other sensory neurons or between sensory and motor neurons. Impulses in such neurons will occur, and in turn themselves acquire connexions, only in so far as they are part of the following of other impulses; but once they have acquired such a position in the system of connexions, they will in turn be able to acquire their own following and this will include impulses belonging to the following of all the different other primary impulses of which they form a part.

3.54. In the higher centres there occur undoubtedly a great many impulses which do not uniquely correspond to particular stimulations of sensory receptors but which represent merely common qualities attributed to the primary impulses; these representatives of classes of primary impulses will in turn become the objects of further processes of classification; the classes for which they stand will be further grouped into classes of classes, and this process can be repeated on many successive levels. We need, of course, not assume that these 'levels' are clearly separated or that the same impulse may not form part of the following of several other impulses which belong to different 'levels'.

3.55. The process of classification which we are considering is therefore 'multiple' not only in the two senses which we have discussed before (2.39–2.43), but also in a third sense: it can take place on many successive levels or stages, and any one of the various classes in which an impulse may be included may in turn become the object of further classification. This third sense in which this process of classification may be multiple must not be confused with the second (2.41); the latter refers to the case where groups of simultaneously occurring impulses $(a, b, c,)$, (e, f, g), (i, k, l), which, when they occur as groups, are as groups treated as members of the same class of groups. In the third sense multiple classification refers to the class A (of which individual impulses a, b, c, or groups of impulses (a, b, c), (e, f, g), etc. may be members) and the class B (of which the impulses m, n, o, or the groups of impulses (m, b, o), (p, q, r), etc. are members) and the similar class C, or rather to the 'symbols' representing the classes A, B, and C, which, by the common following they acquire, become members of a class of a higher order.

3.56. These different forms of multiple classification which it is necessary to distinguish conceptually, will, of course, occur in various combinations, and we obtain thus possibilities of classification of (or discrimination between) the different individual impulses and groups of impulses which are practically unlimited. The consequent differences in the influence which different impulses will exercise on the whole course of the nervous processes, varying from identity through various degrees of similarity to complete difference, would be adequate for building up an extremely complex system of relations among the millions of impulses.

3.57. The word classification scarcely conveys an adequate idea of the almost infinite wealth of variety and gradation of the discriminations which can be performed by such an apparatus. Since it is not merely a question of a particular impulse either belonging or not belonging to a particular class, but also of its belonging to it more or less 'strongly' (according as the connexions with the classifying impulses are 'effective' or merely 'potential', and therefore in the latter case requiring more or less support in order to become 'effective'—3.24 and 3.37), it would be more appropriate to describe these complex processes by some such term as 'evaluation'. We shall occasionally employ this latter term in the place of 'classification' in order to stress that the process is capable of making distinctions of degree as well as distinctions of kind.

3.58. The combination of the different kinds of multiple classification opens up the possibility of a still further organization of the order of the impulses, because through it the differences of the positions occupied in the whole system of classification by the impulses belonging to different classes may themselves become the object of classification and thereby acquire distinct qualities of their own.

3.59. It has been pointed out before (3.17) that it is a somewhat misleading and artificial approach to trace the effects of a single afferent impulse as if it ever occurred in isolation and as if its position were to be determined in an otherwise quiescent system; and that it is doubtful whether such a single isolated impulse, even if it ever occurred, could produce a sensory quality. It is probable that only groups of impulses as such can acquire that distinct position in the whole system which we call its quality. There exists, moreover, a good deal of physiological evidence which makes it probable that it is the so-called 'gradients' between different impulses rather than the individual impulses which are the significant features[1]. We must, therefore, consider more fully this case where not a single impulse but only certain groups of impulses as groups acquire a distinct following of their own (the second kind of multiple classification) and where in consequence the specific following determining a class of groups of impulses will be evoked only if the whole of a group belonging to this class occurs.

3.60. The constituents of a following that will appear only if certain impulses forming a group occur together must be connected with the individual impulses forming the group by what we have called 'potential' connexions. The simplest instance of such a position would be provided by several primary impulses which possess potential connexions with the same other secondary impulse which will be called forth only if all the primary impulses forming the group occur at the same time (or in rapid succession). Several different groups of such individual impulses may evidently thus become connected with the same symbolic impulse (or following of impulses) which will then stand for a class of such groups of impulses. And the symbolic or secondary impulse (or following of such impulse) which stands for any one of this class of groups of impulses (performing thus the second kind of multiple classification) may then (by the third, or relay type, of multiple classification) become itself a member of some new and higher class of impulses representing classes of groups of impulses. This higher class will then be represented by impulses which are symbols of classes of symbols, and so on.

3.61. As a result of such combinations of the different kinds of multiple classification it is evidently possible that the simultaneous occurrence of members of several different pairs (or groups) of different classes of impulses, will be classed as similar events or, we might say, as different events related similarly to each other. Since in such a case the same classifying impulse or impulses will be evoked by different pairs (or groups) of impulses which separately do not belong to the same class, it is legitimate to speak here of a classification of the difference (or relations) between classes of the first kind.

3.62. In order to bring out distinctly the meaning of such a classification of the difference (or relation) between different

[1]Cf. E. D. Adrian, 1947, p. 82, and especially V. von Weizsaecker, 1947, *passim*.

classes, it will be useful to consider the different meanings of the expressions 'respond differently to different impulses', 'show the same difference in the response to different pairs of impulses', and 'respond to a difference between impulses'. In these expressions 'respond', of course, does not necessarily refer to any peripheral response of the organism but to the symbolic or classifying responses in the central nervous system. 'To respond differently to different impulses' then corresponds to what we have called simple classification. 'To show the same difference in the response to impulses of different pairs which in other respects are classified as equal' means that although, e.g., the impulses a and b in most respects belong to the same classes and similarly the impulses e and f also belong in most respects to the same other classes, there is at least one reaction which a and e and another which b and f have in common. 'To react to a difference', finally, means that any member of the class A occurring with any member of class B will produce the same response. If this same classifying response is also evoked by the occurrence of any member of class E together with any member of class F, and by the occurrence of any member of class K together with any member of class L, we can say that the differences or relations between the classes (or qualities) A and B, E and F, and K and L are the same.

3.63. The impulses which in this manner come to stand for, or to represent, particular classes of relations between other impulses will in turn also acquire their own following and thereby obtain their own distinct functional significance: the qualities represented by their common following would attach to the relations between the primary impulses rather than to those impulses themselves. Or, to express the same idea differently, the various kinds of relations between different impulses may themselves become differentiated from each other and thus become capable of forming the starting points of distinct chains of further impulses.

3.64. The relations between impulses or classes of impulses may thus be ordered as a system, or be classified, in the same way and by the same kind of process by which the individual impulses or groups of impulses are arranged in an order. It is in fact only at this point that, strictly speaking, we are entitled to speak of different relations between the impulses (see 1.56–1.61 and 2.20). It will now be seen how, as a result of the hierarchical organization of the connexions between different impulses, the one kind of 'relation' from which we started (namely the causal connexion between the impulses) can be used to build up complex structures with regard to which it is legitimate to speak of different kinds of relations existing between the various elements.

3.65. This process, by which the relations on which the classification of the primary impulses is based, become in turn the object of classificatory processes, can evidently be repeated on many levels. Not only relations between impulses, but relations between relations between impulses, and so forth, may all acquire their distinct following and in consequence become capable of forming the starting point for distinct further proesses.

3.66. The complexity of the order which can be built up by means of this variety of relations is for all practical purposes unlimited. Given the number of separate neurons in the higher nervous centres and the number of the possible connexions between them, the problem is not one of the limitation of the number of possible differences between their respective positions in the whole system, but rather the inadequacy of our mind to follow out the full degree of complexity of the order which can thus be determined. It seems indeed that any conceivable order or structure of relationships could be reproduced within such a system.

3.67. The differences in the functional significance or in the 'quality' which different groups of impulses may acquire as groups, and which may be independent of the functional significance which the individual impulses forming these groups possess if they occur singly, is thus a problem of the same character as that of the functional differentiation of the individual impulses, and can be answered by recourse to the same principle. But although the processes which bring about these differentiations are in principle independent of each other, and while it is even possible that only the classification of the groups and never that of individual impulses is the significant phenomenon, these classifications on different levels will, of course, interact with each other.

3.68. There will thus exist as much justification for saying that the capacity of the individual impulses to combine with others into groups possessing distinct functional significance will contribute to the distinct character which these impulses possess individually, as there is for saying that the latter will contribute to the distinctive following possessed by the group as group. Neither of these two aspects of what is a single process can in any sense be regarded as more fundamental. Both contribute in the same way to the organization of the whole system of sensory qualities; and it is the whole complex order thus produced which determines the characteristic position within this order of individual impulses as well as of groups of impulses.

3.69.   The fact that chains of further processes ('associations') can be evoked not only by the 'elementary' sensory qualities (which were supposed to correspond to the occurrence of particular primary impulses), but also by certain 'abstract' attributes of different groups of sensations (such as figures, tunes, rhythms, or abstract concepts), has usually been regarded as an insurmountable obstacle to any physiological explanation of mental processes.[1] For the approach followed here no such difficulty arises: the problem of the equivalence of 'similar' complexes of stimuli is not different in principle from the problem why the same associations should become attached to different impulses which correspond to the same 'elementary' qualities. The problem of equivalence in both these instances is basically the same and can be solved by the application of the same general principle of explanation.

3.70.   Once a given impulse has acquired a definite following in common with other impulses, any new connexion which it acquires will become attached also to the impulses of its following and will be evoked, therefore, also by the other primary impulses with which it shares part of its following, although those other primary impulses may never have occurred at the same time with those others with which they become in this indirect manner associated. If all the different qualities which different impulses have in common are represented by certain symbolic impulses standing for these qualities and included in the following of all the impulses possessing that quality, there is no difficulty about the manner in which associations will become attached to such common qualities of different impulses rather than to the individual impulses. The phenomena of transfer and generalization of

[1]Cf., e.g., G. F. Stout, 1915, p. 88, and E. D. Adrian, 1947, p. 82; 'The nervous system reacts to relations between stimuli and performs the appropriate task with any part of the motor system that is available. We cannot represent it as a series of machines for operating upon the map of events unless we add a number of devices to make good this fundamental difference. On the sensory side there must be something to abstract the significant element of a pattern and on the motor side something to do just the reverse, to convert the abstraction into concrete movement.'

G   75

learning (1.50) are a direct consequence of the fact that identical mental attributes are represented by identical physiological impulses.

6. THE UNIVERSAL CHARACTER OF THE PROCESS OF CLASSIFI-
CATION: GESTALT PHENOMENA AND ABSTRACT CONCEPTS

3.71   The fact that relations between the parts of the total sensory situation, which individually may be quite unlike each other, may yet be recognized as similar, of course, is the most general aspect of the problem of gestalt. But while the significance of the phenomenon has come to be generally appreciated mainly as the result of the work of the gestalt school, it is by now recognized by practically all schools of psychology. That in perception we do not merely add together given sensory elements, and that complex perceptions possess attributes which cannot be derived from the discernible attributes of the separate parts, is one of the conclusions most strongly emphasized by practically all recent developments in psychology.

3.72.   As we have seen, it is, in fact, no more difficult to explain why different impulses caused by different combinations of stimuli, which singly would occupy altogether different positions in the whole system of relations, should as combinations occupy similar positions in that system, than why different single impulses produced by different physical stimuli should acquire the same or a similar functional significance. That the problem of gestalt perception was singled out as a special problem was largely due to the fact that it was still widely believed that the 'elementary' sensory qualities were somehow originally, and in a manner either not requiring or not capable of explanation, attached to the elementary nervous impulses. The fight which even before the rise of the gestalt school some psychologists had conducted against the 'mosaic psychology', which conceived the more complex phenomena as built up from mental elements corresponding to the physiological elements,[1] was, however, bound to be unsuccessful so long as the purely relative character of *all* sensory qualities was not recognized.

[1]See W. McDougall, 1923, p. x, where James Ward, F. H. Bradley, Dawes Hicks and G. F. Stout are mentioned as protagonists of the fight against the 'mosaic psychology'. A similar list of German writers of that period could be given.

3.73. With regard to the more complex sensory phenomena our theory leads indeed to conclusions very similar to those of the gestalt school. This, however, is so because our approach leads us to raise with regard to *all* sensory qualities, even those presumed to be the most 'elementary', the same question which the gestalt school raised with regard to configurations. Once we are led to account even for what used to be regarded as 'simple' or 'elementary sensory qualities by the principles outlined here, gestalt phenomena and 'abstractions' do not raise any fundamentally new or different problem.

3.74. As a result of the work of the gestalt school the view has now become widely accepted that sensory qualities must not be regarded as atomic fact but should be conceived as determined by the 'organization of the field'. It may be suggested that the theory of the determination of sensory qualities here developed gives this somewhat vague conception of the 'organization of the field' a precise meaning; and, at the same time, that it takes this whole approach some steps further by making it clear, firstly, that the 'organization of the field' is based on, and is in principle capable of explanation in terms of, causal connexions between physiological impulses; and, secondly, that this organization of the field is not additional to the qualities of any kind of atomic sensations (as most of the discussion of 'perceptual organization' still implies), but that it is the structure of that field which determines the peculiar functional significance of the individual impulse, or groups of impulses, which we know as their sensory qualities.

3.75. The conception of the 'organized field' is usually applied to the system of qualities belonging to one particular sense or modality. For our purposes it will be necessary to interpret its meaning more widely and to include in the conception not only the relations between the different qualities belonging to the same modality, but also the relations which exist between the qualities belonging to different modalities (1.56–1.67). The fact that the whole system of sensory qualities must in this sense be regarded as one organized field need not prevent us, however, from occasionally speaking of different fields as sub-systems of the more comprehensive system—sub-systems within which the elements are differentiated by a more dense and complex system of relations.

3.76. In treating the so-called elementary sensations and the more complex sensory phenomena as instances of the same process, and, therefore, as being capable of being explained by recourse to the same principle, we arrive (again in agreement with the views of the gestalt school) at the conclusion that there is no substantial difference between the acts of 'sensation' and of 'perception': both appear as essentially similar and, as we shall see later, they constitute merely different stages in an even more comprehensive range of processes, all of which can be interpreted as acts of classification (or evaluation) performed by the central nervous system. We shall therefore henceforth use the terms 'perception' and 'perceiving' in their popular meaning in which they include the experiencing of 'elementary' sensory qualities as well as the perception of shapes, objects, etc.[1]

3.77. It will be shown later (6.44–6.50) that the principle used to explain these phenomena applies also to the so-called 'higher' mental processes such as the formation of abstract concepts and conceptual thought. With regard to those we are, of course, more familiar with the interpretation as processes of classification in which classes of events, or classes of such classes, interact in a complex manner. It should be noted, however, that if what are called abstractions are most easily accounted for as classes of classes, etc., this does not mean that they must always be secondary, in the sense of being derived from previous conscious experience. The perception of an abstract feature of a situation may in some measure be independent of the perception of the 'concrete' elements of which that situation may seem to be made up (6.40).

3.78. The processes of classification and re-classification on successive levels, and the 'higher' mental processes corresponding to them, will have to be considered further (Chapters V and VI) in connexion with the whole process of the building up of the system of connexions as a whole. Before we can turn to this, however, we must consider another source of classification which, in consequence of the simplifying assumptions made, we have so far disregarded.

---

[1]On the apparent conflict with the views of the gestalt school in which we have been led by explaining the formations of gestalt qualities by a sort of experience see 5.16 below.

CHAPTER IV

SENSATION AND BEHAVIOUR

1. SENSATION AND THE ORGANISM

4.1. In the preceding chapters the apparatus for the classification of impulses has been represented as if it were a neutral, self-contained, and completely centralized system which passively registered the simultaneous occurrence of impulses set up by external stimuli and thus came to reflect the significance which the stimuli possessed in the environment of this system. Such a passive apparatus of registration is conceivable, and to consider it served to bring out the general principle of our theory. But it would be something very different from the sort of apparatus which the nervous system constitutes. While it would register the significance of the stimuli in the environment, it would not indicate the special significance which they possess for the living organism of which that apparatus forms a part.

4.2. The exclusive concentration on the order that might be created by the establishment of the connexions between sensory impulses only has been adopted quite deliberately (3.18). It was intended to emphasize one aspect of the more complete picture which, under the influence of behaviourism, has been somewhat neglected during the last generation. The emphasis which was placed during that period on the observable peripheral responses brought it about that the rôle played by the higher nervous centres has been largely disregarded, and that the whole relation between stimulus and response has often been treated as if the higher centres did not exist. In the preceding chapters we have gone to the other extreme and practically disregarded everything except the central effects of any sensory impulse. This temporary disregard of the fact that the nervous system operates within a living and acting organism, which in some measure is capable of adaptive and regulative behaviour apart from control by the higher nervous

79

centres, must now be corrected by an explicit consideration of these facts.

4.3. In the present chapter, therefore, we shall have to examine not only the effects of the sensory on the motor processes, but also have to give much greater attention than we have yet done to the sensory impulses set up by the various processes in the body, that is, to the registration of stimuli which originate in what has appropriately been called the *milieu intérieur*, the internal environment, within which the central nervous system functions. Of the latter we shall have to conceive as a sort of apparatus of control superimposed upon a living whole rather than as a self-contained and fully centralized structure of its own.

4.4. In turning to these problems we are entering a field in which the very attitudes which during the past generation have led to a comparative neglect of our main problem have led to great progress and the accumulation of a wealth of new knowledge. We have nothing to add to this and cannot even hope to give the barest outline of all the relevant facts which a more systematic survey of the field would have to take into account. The sole purpose of this chapter is to show how our theory of the determination of sensory qualities fits into the picture of the 'integrated action of the nervous system' which is gradually emerging.

4.5. At the same time it should be pointed out, however, that in one respect in which the task which we are undertaking is most in need of a solid foundation, theoretical biology is only just beginning to provide the needed theoretical tools and concepts. An adequate account of the highly purposive character of the action of the central nervous system would require as its foundation a more generally accepted biological theory of the nature of adaptive and purposive processes than is yet available.

4.6. The consideration of the inter-relations between sensory and motor processes will also make it necessary to take more explicit notice of the hierarchical order of the central nervous system. We shall see that the organization of all connexions between sensory and motor processes on many superimposed levels, and the corresponding existence of a hierarchy of centres of increasing comprehensiveness, is of the greatest importance for the understanding of the sensory order.

4.7. In some measure connected with this hierarchical order of the nervous system is the distinction between the phylogenetic and the ontogenetic aspects of the processes in question, or between those connexions which are inherited and those which are acquired by the individual. There is, however, not a great deal which, in the present state of our knowledge, can be said on this question; we shall on the whole have to continue to disregard this distinction and to represent the process of the building up of the sensory order as if it took place in the course of the life of the individual.

4.8.   The relation between sensory and motor processes which we shall have to consider is a double one: we shall have to consider both how various complexes of sensory impulses will influence behaviour, and how in turn the motor responses will influence sensory discrimination. The latter question will make necessary some consideration of the interoceptive and proprioceptive impulses, i.e., those impulses which record not external stimuli but various states of different parts of the organism.

4.9.   A more systematic discussion of the connexion between the sensory and the motor apparatus would also have to include an examination of the manner in which the efferent (or motor) impulses are themselves ordered so as to produce certain co-ordinated patterns of movement, and of how the bundles of efferent impulses interact with the proprioceptive afferent impulses by which the resulting movements are recorded at the centres. In this respect we can, however, attempt no more than the barest sketch which must serve as an indication of the sort of problems which a fuller elaboration of our theory would have to answer.

4.10.   In examining the significance of the proprioceptive impulses we shall have briefly to consider not only the effects which the impulses recording postures and movements accompanying perceptions have on sensory discrimination, and the rôle played in this connexion by the back-reports of responses which are produced by the stimuli at various subcortical levels; but in particular also to examine the significance of the various 'biogenic' impulses which are caused by the vegetative processes of the organism and are closely connected with the various 'urges', 'drives', or 'wants'. The latter are, of course, essential for any explanation of purposive behaviour.

2. EVOLUTION AND THE HIERARCHICAL ORDER OF THE CENTRAL NERVOUS SYSTEM

4.11. The continued existence of those complex structures which we call organisms is made possible by their capacity of responding to certain external influences by such changes in their structure or activity as are required to maintain or restore the balance necessary for their persistence. This involves, even in the most primitive organisms, some capacity of discriminating responses to different physical stimuli, and perhaps even some capacity of 'learning',[1] although we know very little about the nature of such individual learning (as distinguished from the process of hereditary selection of such individuals as show appropriate adjustments).

4.12. The fact that an organism will respond differently to different external forces acting upon it is, of course, not peculiar to organisms. It would be merely an instance of different causes producing different effects. The peculiar problems presented by organisms appear only where they respond to particular stimuli in the manner which will secure their continued existence, and in so far as they develop specific organs which enable them not only to discriminate between different stimuli, but to react differently to the same stimuli if they appear in different combinations with other stimuli or when the organism itself is in different states.

4.13. It is perhaps worth stressing that the problem of purposive adjustment of organisms to changes arises long before the problem of its purposive behaviour with regard to external objects. The question of what determines (or what is meant by) purposiveness is in the last instance really the same question as that of what ensures the continued existence of the organism. It arises as much in connexion with the normal functioning and growth of the organism, the processes of metabolism and the replacement of damaged parts, as in connexion with those movements of the organism which we commonly describe as behaviour.

4.14. It has already been suggested that in a certain sense any attempt to explain the highly complex kind of purposive action made possible by a developed central nervous system may be premature so long as we do not possess a fully adequate biological theory of the comparatively simpler kind of purposive functioning. Many of the problems often regarded as peculiar to mental

[1]See H. S. Jennings, 1906.

phenomena in fact arise already at a much earlier stage, where there can be no question yet of that complex order that is shown in the response to external stimuli which we have described as mind. It cannot be our task here to restate the present position of biological theory with regard to these problems, and we must content ourselves to refer in this connexion to W. B. Cannon's concept of homeostasis and its development by other authors, and especially to the more recent and most promising work of L. von Bertalanffy. His theory of 'open systems' in a steady state (*Fliessgleichgewicht*) in which 'equifinality' prevails because the equilibrium that will be reached will be in some measure be independent of the initial conditions, seems to provide the most helpful contribution to this problem.[1] Any further comments we shall have to make with particular reference to the problem of purposiveness will be reserved to the next chapter (5.63–5.76).

4.15. Here we are not directly concerned with the regulative functions of the organism other than those which are effected through the central nervous system. It is merely necessary to remain aware throughout that this system functions within an organism which independently of the former is capable of some adaptive and purposive responses to external causes, responses which are brought about through a system of neuro-chemical regulation. Our task begins essentially where the somatic nervous system makes possible discriminatory responses to a great variety of combinations of stimuli, and particularly where learning becomes the dominant factor.

4.16. The mere fact that the organisms in the course of their evolution develop specific receptor organs which are sensitive only to fairly narrow ranges of stimuli must not be confused with the development of a sensory order and of distinct sensory qualities. It is to be assumed that this development goes hand in hand with the acquisition of different motor responses to the different stimuli. But even an organism which had developed distinct receptors sensitive to all those various stimuli which in a highly developed central nervous system produce different sensory qualities, for that reason could not yet be said to discriminate in terms of a system of sensory qualities similar to that familiar to us.

4.17. While such an organism would perform the simplest kind

[1]W. B. Cannon, 1932; L. V. Bertalanffy, 1942, 1949. See also J. H. Woodger, 1929.

of classification we have discussed (2.35–2.38), it would still be unable to perform the multiple classifications which alone give rise to the system of sensory qualities. The different stimuli which evoke the different sensory qualities would, if occurring in isolation, all produce different effects, but these different effects would not differ from each other in the specific manner in which the sensory qualities differ from each other.

4.18.   The essential characteristic of the order of sensory qualities is that, within that order, each stimulus or group of stimuli does not possess a unique significance represented by the particular response, but that they are given different significance if they occur in combination with, or are evaluated in the light of, an infinite variety of other stimuli which may originate from the external world or from the organism itself.

4.19.   Such an order implies that any impulse recording a particular stimulus is connected not merely with one particular motor response, but that some apparatus exists by which the effects of any impulse are adjusted to, and integrated with, the effects of other impulses proceeding within the central nervous system at the same time. In other words, the various sensory impulses, whose effects are thus to be adjusted to each other, must in some manner be brought together before the effect of their joint action is decided.

4.20.   This does not mean that individual afferent impulses, or groups of such impulses, might not also be uniquely connected at low levels with certain efferent impulses so that, as soon as the former occur, a particular movement is produced. Such a relation would correspond to the ideal simple reflex arc of traditional theory. In the present context such reflex responses are important mainly because of the proprioceptive impulses by which they in turn will be recorded in the higher centres. The original exteroceptive impulse which sets up such a reflex will in consequence arrive at the centres to which it is conducted already accompanied (or rapidly followed) by a report of the spontaneous response of the organism to the external stimulus. The impulse recording the external stimulus is thus already 'marked' as meaning (involving) a certain kind of response.

4.21.   This sort of relationship we must suppose to recur in relays, or at many successive stages: the initial sensory impulse at the first stage, the spinal cord, setting up both a motor response and a further afferent impulse proceeding on to the higher centres. At the next level it will arrive together with the report of the motor responses it has produced at the lower level, and with other sensory impulses recording other peripheral stimuli, which may be similarly accompanied by the reports of the reflex responses which they have set up at lower levels. At this stage this particular combination of signals may again produce a distinct motor response, so that at the next higher level the bundle of impulses arriving there will include reports of responses which take already a wider range of exteroceptive impulses into account. And as we ascend to higher and higher levels, both the comprehensiveness of the range of external stimuli which are taken into account in any response, and the number of responses effected at lower levels and reported back to these higher levels will constantly increase.

4.22. It is not difficult to see how such an arrangement for the mutual adjustment of the responses to different simultaneously occurring stimuli is made necessary by the development of specific receptors for different kinds of stimuli. So long as the whole organism was merely susceptible to irritation by a wide range of stimuli, and was capable only of a few simple responses, such as contraction and expansion, no special apparatus for co-ordinating the responses to different stimuli was required. But as soon as specific responses became attached to particular classes of stimuli, the mutual adjustment of these responses according to the significance of the particular combination of stimuli became necessary.

4.23. Between the case where specific responses are uniquely attached to particular stimuli, and the case where all responses are decided in view of all the stimuli, there is, of course, an enormous range of intermediate possibilities. Nor need in a given organism only either the one or the other type of arrangement exist. Some other stimuli will be more likely to affect the appropriateness of a given response to one particular stimulus than will be true of others, and there will be more need—or it may be easier for the organism to provide—for the mutual adjustment between those than between others. We must probably assume that, in the course of evolution, the original direct connexions between particular stimuli and particular responses are being preserved, but that control mechanisms are being superimposed capable of inhibiting or modifying these direct responses when they are inappropriate in view of other simultaneously acting stimuli.

4.24. Parallel with this progressively more complex evaluation of the stimuli in the light of an ever more comprehensive collection of other stimuli, a similar organization will operate on the motor side: instead of simple movements of particular muscles, more and more complex patterns of behaviour will be evoked as a whole; and the groups of impulses which evoke this pattern of movements are probably evoked as groups by a few central impulses which 'stand for' the whole pattern (4.48–4.51).

4.25. The sketchy manner in which these questions must be treated here ought not to give the impression that these relations are simple. It is neither to be assumed that comparatively simple patterns of stimulation will normally produce comparatively simple responses, and that the integrative action of the higher centres will operate only when more complicated stimulation patterns are involved; nor that the motor responses are built up in a simple additive manner from the effects of individual impulses producing the movement of individual muscles. It is very likely that, just as more than one afferent impulse will generally be required to produce anything like a 'simple' (or sharply discriminated) sensation, so a single efferent impulse will as a rule produce somewhat diffuse movements and only the overlapping of many such impulses will produce a clearly differentiated movement of the separate muscles.[1]

4.26. The increasing differentiation of the different stimuli from each other, and the accompanying increased variability and complexity of the responses to any subgroups of stimuli, involves, as we have seen, that impulses representative of these stimuli are brought together so that they can act upon each other in a manner which reproduces their significant relations. The more comprehensive this adjustment is, the more elaborate must be the centres set aside from, and capable of overruling the effects of, the more direct connexions between stimulus and response.

3. FROM SPECIFIC REFLEX TO GENERALIZED EVALUATION

4.27. It is doubtful whether the ideal simple reflex arc, where the impulse from a single afferent fibre is transmitted to a single efferent fibre, is of any importance, and even whether it occurs at all. But between it as the one extreme ideal type and the other

[1] V. von Weizsaecker, 1940, p. 48.

extreme of the 'voluntary' or 'conscious' responses there exists probably a continuous range of connexions between stimuli and responsies of intermediate types in which processes of classification take place which are more or less analogous to those which determine the system of sensory qualities. The simplest of these intermediate cases which is of interest is that in which a particular motor response becomes attached to every one of a group of sensory impulses, so that any one of the latter will be transmitted to the motor fibre and produce the response in question. This represents the simplest possible instance where we can speak of a classification of the stimuli.

4.28. The operation of this simple kind of classification is familiar from the experiments with conditioned reflexes and from the phenomenon known as generalization. It has been found tha after a conditioned response has been developed to one stimulus, other 'similar' stimuli may also elicit the same response.[1] In these instances the grouping of certain impulses as similar has the effect that further impulses which become connected to some of these impulses become also attached to the other impulses forming the group.

4.29. A higher degree of selection or classification is reached when several responses are alternatively connected with each of a given group of sensory impulses, so that which of these responses will be elicited by a particular stimulus depends on which of a number of other sensory impulses occur at the same time with the former. Which response will be produced in this case by any sensory stimulus will depend on what other stimuli accompany it, and any particular sensory impulse may in some context produce results which are similar to those produced by others, and in other contexts produce results which are different.

4.30. On the lower levels on which connexions of this still relatively simple or quasi-reflex type prevail, there will thus already exist some sort of qualitative ordering or discrimination.

[1] E. R. Hilgard and D. G. Marquis, 1940, p. 46.—The conditioned reflex is usually represented as a comparatively recent discovery, but the bare facts have been known for a very long time and were already described by M de Montaigne in the chapter on "The Force of Imagination" of his *Essais* (1580, Book I, Chapter 20). He there describes the case of a man who, after he had for a long time regularly tested with his hand the temperature of the water prepared for an enema, found that the actual injection had become unnecessary because his taking up the appropriate posture together with his placing the hand in the water already produced the desired effect.

But it will be greatly limited in two respects: it will involve selective responses to the stimulation of particular sensory receptors for only a very few kinds of responses: and it will select between this limited number of possible responses only with regard to a limited number of simultaneously occurring other stimuli.

4.31. While on these lower levels the discrimination may thus be fairly detailed in so far as it refers to particular responses or functions, it will be specific in the sense that it will be effective only with regard to a particular group of responses, and take into account only a comparatively small range of stimuli. In the famous case of the decapitated frog which is still capable of wiping a drop of acid from its back, the signal evoked by the drop of acid will be sufficiently discriminated as regards location to guide the movement of the leg. But the localization of the stimulus which this proves will probably be specific in the sense of being effective only with respect to this particular response.

4.32. Such limited classification may be effected in subcentres which offer the opportunity for connexions among a limited number of sensory and motor fibres. As the impulse is transmitted to higher and more comprehensive centres, there will arise opportunities for more extensive connexions, and with them will appear the possibility of a more complex discrimination both with regard to the range of different responses and to the variety of stimuli which will contribute to the decision which of the potential responses will take place.

4.33. The increasing opportunities for connexions between fibres carrying sensory impulses from different parts of the body, and the correspondingly increased comprehensiveness of the net of connexions which can be formed on the higher levels, does mean neither that at these higher centres the individual stimuli must always be represented by individual impulses as they are on the lower levels, nor that the lower level connexions are confined to impulses belonging to the same sense modality.[1] It means merely

[1] V. von Weizsaecker, 1940, p. 55, points out that, e.g., to the more than four million points on the skin below the neck which produce distinct sensations correspond at most one half million fibres conducting the impulses set up by these stimuli beyond the spinal level. That in spite of this these individual stimuli can be distinguished is presumably due to the fact that with the report of any stimulus acting on the skin there will also arrive reports of the low level reflexes caused by them, reflexes which may be different for stimuli whose direct report arrives in the brain through the same last common path.

that at the lower levels only such other sensory impulses will generally be able to modify the response to a particular impulse as are most immediately relevant to the interpretation of (or most frequently associated with) the particular stimulus; while at the higher levels a wider range of less immediately significant other factors will have an opportunity of modifying the result. Similarly, the increasing comprehensiveness of the connexions possible at the higher levels *need* not mean the possibility of connexions with a larger *number* of individual impulses, but may mean merely the possibility of connexions with impulses representing a greater variety of stimuli.

4.34. The responses to any given stimulus thus become at the higher levels more and more liable to be modified by the influence of impulses from other sources. The continuous range of connexions between the simple reflex and conscious action thus becomes one in which an ever-increasing number of different stimuli contribute jointly in determining the response. Even though we are directly familiar only with the classification of stimuli which lead to behaviour that is conscious, and in which this comprehensiveness of the stimuli taken into account has presumably reached the highest degree, at least a great part of observable behaviour is probably guided by processes which are intermediate between this and reflex action.

4. PROPRIOCEPTION OF LOW LEVEL RESPONSES

4.35. The fact that the sensory impulses may evoke responses on many successive levels has great influence on the manner in which they will be discriminated at the higher levels. In so far as sensory impulses evoke such responses at lower levels, they will arrive at the higher levels accompanied by the proprioceptive impulses recording those responses. The higher centres will in consequence at any one time receive reports not only of given external stimuli but also of the body's spontaneous reaction to those stimuli. The effect of a bright light will not be only a visual impulse but also an impulse reporting the contraction of the pupil, etc. So far as the higher centres are concerned, the self-moving organism must indeed be regarded as part of the environment in which they live.

4.36. Since as a result of the excessive stress placed by the behaviourists on the peripheral responses, certain misconceptions about their significance are still prevalent, it will be necessary to consider with some care the rôle which such peripheral movements can play in the structure of nervous action. The first point which requires emphasis is that peripheral events, in order to influence further central nervous processes, must be reported back to the centres in which these processes take place. It will therefore be neither the resulting movements as such, nor the efferent motor impulses, but the proprioceptive impulses recording these movements which affect the further neural processes. (The theoretical possibility that part of every efferent neural impulse may, as it were, be branched off before it leaves the centre from which it originates, so as to represent there the resulting movement, can be disregarded because there appears to exist no evidence for this.)

4.37. This means not only that, even where distinct motor responses to the individual stimulus take place, it will still be the (proprioceptive) sensory impulses and not the motor impulses themselves which are important for our purposes, but also that, once a certain peripheral response has become the regular effect of any group of stimuli, it will no longer need actually to occur, since the reports of its occurrence will be associatively evoked by the original stimulus. The emphasis placed by the behaviourists on actual movements, and their efforts to discover at least traces of such movements in the form of 'implicit speech' and the like, were thus misplaced. They are not required and the establishment of their existence would not help to answer the problem of what, for instance, constitutes thought.[1]

4.38. Nevertheless, it is true that the sensory order with which we are concerned is both a result and a cause of the motor activities of the body. Behaviour has to be seen in a double rôle: it is both input and output of the activities of the higher nervous centres. The actions which take place independently of the higher centres help to create the order of the sensory impulses arriving at that centre, while the actions directed from that centre are determined by that order.

4.39. The evaluations of sensory impulses arriving at the highest centres may be compared to the appreciation of the events on the road observed by a person who is being driven in a car, or to the judgements of the pilot of an aeroplane which is being steered by an automatic pilot. In these instances different observed events will

[1]See C. T. Morgan, 1943, p. 476.

lead the passenger of the car, or the pilot of the plane, to expect certain responses of the car or the plane, and those events will come to 'mean' for the person particular kinds of responses of the vehicle, just as certain kinds of stimuli mean certain spontaneous responses of the body. The sight of an oncoming car will come to mean the sensation of the car in which the person rides drawing to the right, and the sight of a red traffic light will mean the feeling of the car slowing down. Very soon what will actually be noticed will no longer be that normal response, but only its absence if it fails to occur.

4.40. The position of the highest centres in this respect is somewhat like that of the commander of an army (or of the head of any other hierarchical organization), who knows that his subordinates will respond to various events in a particular manner, and who will often recognize the character of what has happened as much from the response of his subordinates as from direct observation. It will also be similar in the sense that, so long as the decision taken by his subordinates in the light of their limited but perhaps more detailed observation seems appropriate in view of his more comprehensive knowledge, he will not need to interfere; and that only if something known only to him but not to his subordinates makes those normal responses inappropriate will he have to overrule their decisions by issuing special orders.

4.41. In the same manner as, for instance, the captain of a battleship may sometimes recognize the nature of an observed object less from his direct perception of it than from the responses of his ship, so the brain often may get a direct report merely about the action of one of a large class of stimuli and yet be able to recognize its character from the almost simultaneous reports of the responses of the body directed by lower levels. At the same time these responses of the lower centres to particular stimuli, of which the higher centres have no reports, may be governed by general 'directives' issued by the higher centres. (We shall see presently that this 'set' of the whole organism which determines what the response to a particular stimulus shall be, may in turn be determined either by processes in the highest centre or be the result of subcortical regulation).

4.42. So far as the higher centres are concerned, a given combination of external stimuli will not merely mean that such and such other external events are to be expected, but also that certain adjustments of the organism are taking place. The significance of the effect of cold on the skin will not only be that certain action is indicated, but also that certain responses of the body will automatically come about—not merely a report of a single external stimulus but at the same time also of a change in the state of a great part of the body.

4.43. While it is on the whole more likely that responses *via* the lowest centres will be innate for the individual, that is, acquired by the race in the course of evolution, while the responses effected by the higher centres will be largely based on individual experience, this cannot be regarded as a universal rule. Probably some inherited responses are effected on fairly high levels, while some learned responses may, after sufficient repetition, become almost completely automatic and be effected at low levels.

4.44. It should also be noted that the degree of modifiability of the response to a particular stimulus by other simultaneous stimuli need not vary in strict correspondence with the extent to which these responses can be altered by individual experience: an acquired response to a given stimulus may be uniquely determined by that one stimulus, while an inherited response may be capable of considerable variation according to the accompanying circumstance.

5. POSTURES AND MOVEMENTS CONNECTED WITH PERCEPTION

4.45. The first group of motor responses to sensory stimuli which we must consider further are those which assist perception directly and which might almost be described as part of the act of perception. We have already mentioned the classical instance of the kinesthetic sensations connected with the focusing of the eye. The familiar effects of displacing the eyeball or of crossing the fingers on the localization of the sensations affected belong to the same category. It is becoming increasingly clear that these are merely special instances of a very general phenomenon, and that the proprioceptive reports of the body postures and movements designed to help perception serve always as a sort of indispensable background for the proper evaluation of the stimulus.

4.46. Recent investigations of the relation between sensation and movement show that this connexion is even closer than had been commonly supposed and that practically all sensory impulses are evaluated in the light of, or corrected for, simultaneous muscular activities. V. von Weizsaecker, to whom we owe a great deal of knowledge on this question, speaks with considerable justification of a complete 'entwinement' (*Verflechtung*) of sensation and movement.[1] This seems to apply as much to the evaluation of external stimuli in the light of the simultaneous proprioceptive impulses as reciprocally to the evaluation of the latter in the light of the usually accompanying exteroceptive impulses. Stretching my leg downwards means for me that I expect to feel the ground, and stretching my whole body means that I expect it to cool more rapidly than in a crouched position, etc., etc. The proprioceptive impulses thus receive their significance as much from the exteroceptive ones which are associated with them, as the reverse is true.

4.47. Every sensory situation thus means, among other things, that various movements will have such and such effects, and the totality of the simultaneous exteroceptive and proprioceptive impulses forms the background, as it were, against which the individual impulse is evaluated. It might even be said that every single sensory impulse is probably multivalent, capable of producing various different sensations, and that which sensation it will produce will depend on what other impulses occur at the same time.

6. PATTERNS OF MOTOR RESPONSES

4.48. The manner in which the separate motor impulses are coordinated so as to produce complex patterns of behaviour consisting of many simultaneous and successive movements can be considered here only very briefly. We must probably assume that these patterns can be elicited as wholes by a few signals sent out from the higher centres, and that we have thus on the motor side to deal with a phenomenon of 'bundling' which in some respects is the converse of the process of classification on the sensory side. As in the latter case, different complexes of sensory impulses will be represented at the higher centres by a few 'representative'

[1]V. von Weizsaecker, 1940. See also K. Goldstein, 1939; and E. G. Boring, 1942, p. 563: 'In the twentieth century it eventually became apparent that the organism behaves first and feels afterwards, just as James, speaking of emotions, said it does.'

impulses, so that a few central impulses may suffice to evoke bundles of motor impulses producing complex patterns of behaviour. The particular manner in which this behaviour is executed may then be determined by the interplay of motor and sensory impulses at lower levels.

4.49. These behaviour patterns, however, must not be conceived as fixed but as highly variable. Just as at the higher centres it will not be only one particular sensory impulse, but any one of a class of many different combinations of impulses, which will give rise to a particular response, so the motor signal sent out from the higher centres will be for the execution not of one particular pattern of co-ordinated movements but for any one of a class of such patterns. Such a class of patterns will consist of those different combinations of movements which under different conditions will produce a particular result. Which of these patterns will be put into effect will be decided in the light of the whole sensory position.

4.50. At the higher centres the connexions will thus increasingly exist, not between particular stimuli and particular responses, but between classes of stimuli and classes of responses, and between classes of classes of stimuli and classes of classes of responses, etc. The order given by the highest centre in response to a particular situation may thus be of the kind which we have called a general 'directive' for an action of a certain class, and it may be only at lower levels that the appropriate response is selected from the class of behaviour patterns which in different situations may produce the desired result.

4.51. The extent to which behaviour patterns can be adjusted to the sensory situation probably varies with the level which is in control. There is reason to believe that some highly stereotyped or 'mechanical' patterns, such as those of the movements of flying and running, are co-ordinated on a fairly low level and that even at these low levels the execution is constantly controlled and modified by sensory signals from the kinesthetic receptors and the semicircular canals. At higher levels the pattern of movement will be variable to a higher degree.

4.52. We can again not concern ourselves here with the question to what extent behaviour patterns are innate to the individual and how the innate and the learned behaviour pattern interact.[1] There can be little doubt that even fairly complex behaviour

[1]K. Lorenz, 1943, *passim.*

patterns, or rather classes of behaviour patterns from which a selection will be made in the light of the whole sensory situation, are innate and can be elicited by fairly simple stimuli. A well-known instance is the evocation of the maternal behaviour in the rat by definite chemical stimuli.[1]

4.53. The selection of the particular behaviour pattern from the class of such patterns appropriate to the result aimed at, must not be conceived as taking place in one act. The choice of a kind of behaviour pattern and its continued control, modification, and adjustment while it takes place, will be a process in which the various factors act successively to produce the final outcome. It is not as if the whole behaviour pattern were determined upon before any movement takes place, but rather that during the process of execution further adjustments are constantly made to secure the result.

4.54. In connexion with these continuous adjustments, made while the movement proceeds, the interaction between the exteroceptive and the proprioceptive impulses and the operation of the 'feed-back' principle[2] become of special significance. In the first instance, the sensory representation of the environment, and of the possible goal to be achieved in that environment, will evoke a movement pattern generally aimed at the achievement of the goal. But at first the pattern of movement initiated will not be fully successful. The current sensory reports about what is happening will be checked against expectations, and the difference between the two will act as a further stimulus indicating the required corrections. The result of every step in the course of the actions will, as it were, be evaluated against the expected results, and any difference will serve as an indicator of the corrections required.

4.55. In this process the intervention of the highest centres is probably needed only to give the general directions, while the execution and current adjustment is left to the guidance of the lower centres. Once the 'course is set', the deviations will be automatically corrected by the differences between the expected and the effective stimuli acting as the signs which produce the correction. Such responses to a difference between expectations

---

[1] C. T. Morgan, 1943, p. 411.
[2] N. Wiener, 1948 *a* and *b*; W. S. McCulloch, 1948; W. R. Ashley, 1947, 1948, and 1949.

and outcome are merely a special case, on the one hand, of the general principle that a response to any new stimulus is determined by the pre-existing sensory state, and, on the other, of the capacity of the nervous system to respond in a particular manner to certain kinds of relations between impulses rather than merely to particular impulses. Both these characteristics of the higher centres, the predominant importance of the pre-existing excitatory state, and the tendency to respond to differences between expected and realized impulses, will have to be considered further in the next chapter.

7. BIOGENIC NEEDS AND DRIVES

4.56. We still have not yet noticed the prime sources of activity of the organism, namely those changes in its constitution or balance which occur periodically as a result of its normal vegetative processes and which make action by the organism necessary if it is to survive. In consequence of our stress on the sensory organization we have so far treated the whole problem as if it were mainly one of adaptation of the organism to changes originating in the environment. But even more important than the question why the organism will behave differently in different environments is the question why it will at different times behave differently in the same environment.

4.57. Even more than before our discussion must here presuppose a great deal which belongs to theoretical biology and physiology. As has already been pointed out (4.13), there is really no other difference between the problem of the purposive internal functioning of the organism and that of its purposive behaviour towards its environment, than that the latter raises the problem of a comprehensive order of the various external stimuli which determines how in different combinations they will modify each other's effects. This includes the problem of why internal stimuli may bring it about that a given organism will at different times respond differently to the same set of external stimuli.

4.58. What is significant here for our purposes is not so much the precise nature of the physiological processes which determine such states as hunger, thirst, and the like, but what the 'attitudes', 'dispositions', or 'sets' corresponding to these physiological states mean for the responses of the organism towards its environment.

Since these various 'needs' or 'drives' may be produced by visceral, glandular or general metabolic processes, it is convenient to refer to them by the generic term 'biogenic needs'.[1]

4.59. It may be mentioned at once that these 'needs' resulting from the spontaneous vegetative processes of the body are, of course, closely related to, and sometimes practically indistinguishable from, another kind of attitudes or sets such as fear or rage, which, though usually caused by some sensory perception, also consist of a disposition for a certain range or type of actions. The problems which these 'emotions' or 'feelings' raise are thus very similar to those raised by the needs in the narrower sense. It would be difficult to decide whether the sexual urge provoked by a sensory impression is in this sense a 'need' or an 'emotion'. Similarly, appetite may be stimulated by the smell of some delectable food without hunger being present, or a sense of fear caused by bodily processes without any sensory (i.e., exteroceptive) experience inspiring the fear.

4.60. While we shall in the first instance consider the significance of needs and wants proper, and leave to the next sections any more specific comments on emotions, most of what is to be said about needs applies equally to emotions. Both involve not only a disposition of the organism towards a certain class of actions, but also a special receptivity for certain classes of stimuli. As a result of a peculiar state of balance the whole organism comes to 'like' or 'dislike' particular kinds of stimuli. We shall later in connexion with 'attention', (6.26–6.27) have to consider more fully the nature of this state of excitatory preparedness.

4.61. It is perhaps useful to distinguish between the term 'set'[2] as the name for the preparedness of the organism for certain kinds of actions, and the term 'expectancy' for the increased receptivity for certain kinds of stimuli which will elicit the corresponding responses. But such a distinction between the sensory and the motor aspect of what is essentially a relation between a class of stimuli and a class of responses must not lead us to treat them as if they were really separate. The important point is their close connexion, the fact that the organism will be disposed to respond in a particular manner to any one of a class of stimuli.

4.62. At this stage of the exposition it is yet too early to try to

[1]C. T. Morgan, 1943.
[2]See J. J. Gibson, 1941.

show even in outline how such a state of need, which at first may produce merely an aimless increase of motor activity, may become directed towards the purposive search for certain kinds of stimuli, so that the animal searches for food or for a sex partner, which, when found, will produce the consuming activity. This will have to be attempted in the next chapter.

4.63. Our present purpose was merely to show that in addition to the 'objective' significance which the different stimuli will acquire for the organism as the result of their regular association with other stimuli, they will also acquire a special 'subjective' or 'pragmatic' significance through their capacity of satisfying certain needs. The connexions which will give them this significance will operate not only through certain stimuli producing certain actions if the need is present, but also through the need of making the organism search for stimuli of the appropriate kind. This evaluation of stimuli with respect to goals which are determined by the momentary needs will have to be considered further when we examine the general problem of how a representation of the environment enables the organism to act 'purposively' (5.64 ff.).

8. EMOTIONS AND THE JAMES-LANGE THEORY

4.64. The second kind of dispositions, emotions, are dispositions for a type of actions which in the first instance are not made necessary by a primary change in the state of the organism, but which are complexes of responses appropriate to a variety of environmental conditions. Fear and anger, sorrow and joy, are attitudes towards the environment, and particularly towards fellow members of the same species, which may become attached to, and then regularly evoked by, a great many different classes of stimuli.

4.65. This means that a great variety of external events, and also some conditions of the organism itself, may evoke one of several patterns of attitudes or dispositions which, while they last, will affect or 'colour' the perception of, and the responses to, any external event. In the mental order of events, that is in the influence which external stimuli can exercise on further mental processes and on behaviour, these states will occupy positions which in many ways will be similar to those of the sensory qualities: the occurrence of any one of them will be capable of modifying the result of a given sensory situation in the same kind of way in which the appearance of a new sensory experience could do so.

4.66. Emotions may thus be described as 'affective qualities' similar to the sensory qualities and forming part of the same comprehensive order of mental qualities. But they differ, of course, in some respects from the sensory qualities and must be regarded as forming a distinct sub-system of the more comprehensive mental order. The relation between the order of affective qualities and the order of sensory qualities must be conceived as somewhat similar to the relations between the orders of the different sensory modalities which also form sub-systems of the more comprehensive order of all sensory qualities (3.75).

4.67. The most conspicuous difference between the order of the sensory qualities and the order of the affective qualities is that, while the former is organized with spatial relationships as one of its main ordering principles, the affective qualities do not refer to particular points in space. They represent not qualities of particular things but rather a condition of an interval of time as a whole. They will refer not to what is to be expected of an external position but are rather a temporary bias or preference for certain types of responses towards any external situation.

4.68. These important differences between sensory and affective qualities, however, do not alter the fact that the general principle by which they are determined is the same. The similarity of the response to different stimuli will in both instances be determined by the fact that the corresponding different nervous impulses will evoke the same following of other impulses. Similar emotions, as similar sensations, are nervous impulses which evoke the same following and which are therefore functionally equivalent and classified as the same kind of event. The main difference is that within the sensory sub-system of the mental order the classifying connexions will be mainly with other impulses representing sensory stimuli, while in the affective sub-system the classifying connexions will be mainly with impulses representing certain types of behaviour.

4.69. But although the order of affective qualities will constitute a sub-system in the more comprehensive system of qualities (in the sense that the impulses belonging to it will be less closely connected with impulses in other parts of the larger system than they are connected among themselves), this does not prevent this sub-system from contributing to the differences between sensory qualities, and vice versa. By becoming connected with sensory qualities those differently organized qualities can add, as it were, an additional dimension to the order of sensory qualities; and similarly the differences between the different sensory qualities associated to different groups of the latter may assist in enriching the variety of differentiations between the former.

4.70. This account of the determination of the affective qualities of course corresponds very closely to the familiar James-Lange theory of emotions As we said before of the Berkeleyan theory of spatial vision (3.40–3.42), the James-Lange theory also may be regarded as a special case of the theory of mental qualities here outlined. The modifications which are required to make the James-Lange theory fit into our scheme are practically the same as those which we had to make with regard to the rôle which proprioceptive impulses play in determining the perception of space. We shall not regard the actual sensations produced by the various bodily accompaniments of a given stimulus as deter-mining its affective values, but merely the following of physio-logical impulses which record the states of the body and which, in the same manner in which such a following can determine the peculiar functional significance which we know as sensory quali-ties, can also determine affective qualities.

4.71. We do, therefore, not propose to say, with William James, that emotions are 'a set of kinesthetic and organic sensations'. We shall merely contend that the connexions with impulses recording certain connected sets of changes in the general state of the body can give certain central impulses that peculiar position in the whole system of mental events which we know as the different affective qualities.

4.72. The James-Lange theory of emotions (like Berkeley's theory of spatial vision) would thus appear to be justified in its endeavour to reduce the qualitative attributes of those mental events to relations between different impulses which, if fully evaluated, might evoke certain other sensations. Both theories, however, fall short of a real answer to their problem, and in fact merely shift the problem to a different point, because they attempt to explain the quality of one kind of experience by reference to qualities occurring in another kind of experience, which latter they take as not requiring explanation. In so far as they were concerned with only that one kind of mental quality this procedure was inevitable. But if we consistently follow up and generalize the principle underlying those theories, there remain of course no given mental qualities; we are forced to replace the whole system of qualities by a system of relations between initially undifferentiated elements which can be conceived to be isomorphic with the system of qualities which we have to explain.

CHAPTER V

THE STRUCTURE OF THE MENTAL ORDER

I. PRE-SENSORY EXPERIENCE OR 'LINKAGES'

5.1.   In the preceding chapters we have given the general outline of the principle by which a set of neural impulses in principle may become organized in a manner analogous to the familiar order of the mental qualities. We shall now have to try and fill in this outline by a sketch of the process by which this order is formed, and of the general character of that order itself.

5.2.   This account of the formation of the mental order will still have to be extremely schematic, in the sense that we shall not attempt more than a general indication of a possible way in which such an order may be built up, without attempting to show in what manner this will happen in any particular organism. We shall also still have largely to disregard the distinction between the part of this process which takes place in the course of the development of the single individual, and the part which takes place in the course of the development of the species and the results of which will be embedded in the structure of the individual organism when it commences its independent life (or when it reaches maturity).

5.3.   There is at present still very little knowledge available which would enable us to draw such a distinction between the part of the mental order which for the individual is determined by its inherited constitution and the part which for it may be regarded as being of experiential origin.[1] But as we are concerned with the genesis of mind as such, it is comparatively unimportant what for the individual are constitutional and what are experiential factors; indeed, it is at least likely that what for one species or at one developmental stage may be of experiential origin, may in other instances be constitutionally determined. What is important

[1]See, however, the very important contributions to this problem by K. Lorenz, 1943.

for our purposes is that it would appear that the principle which determines the formation of the mental order may operate either in the ontogenetic or in the phylogenetic process. Such an assumption of a general similarity between the kind of processes which take place in the evolution of the species and in that of the individual does, of course, in no way prejudge the issues of the great controversies on the mechanism of evolution.

5.4. For the purposes of the following schematic exposition we shall, therefore, proceed as if at the commencement of the life of the individual the structure of the central nervous system were fully completed before any connexions between neurons corresponding to the simultaneous occurrence of stimuli had been established. This means in effect that we shall disregard the possibility of the transmission from generation to generation of connexions in the higher nervous centres which constitute adaptations to the environment; and that we shall treat a process as if it took place in the development of the single individual which in fact probably occurs to a large extent in the course of the development of the species. This assumption perhaps may be justified in some measure in so far as the highest centres are concerned, but it certainly does not apply to the connexions existing at the lower levels, which form an essential part in the complete process of classification.

5.5. An afferent impulse arriving for the first time at the higher centres of such a system would thus not yet possess any connexions with other such impulses and therefore not yet occupy a definite position in the order of such impulses, or have a distinct functional significance. But since every occurrence of a combination of such impulses will contribute to the gradual formation of a network of connexions of ever-increasing density, every neuron will gradually acquire a more and more clearly defined place in the comprehensive system of such connexions, and with it a distinct functional significance which in a great many ways will differ from that of other impulses.

5.6. In a certain sense it might be said that the qualitative distinctions which will thus be built up between the significance of the different impulses are created by 'experience'. In doing so, however, we should have to be aware that we are using the term 'experience' in a somewhat special sense. Since the impulses between which these first connexions are formed would not yet occupy a place in an order of sensory qualities, and no such order would yet exist, their occurrence could not yet be described as experience in the ordinary meaning of this term. It would not yet represent a mental event but would be a purely physiological event because it would possess none of the attributes which give it a place in a mental or qualitative order.

5.7. The term 'experience' in this connexion is thus somewhat ambiguous and misleading, because it suggests the occurrence of sensory qualities, while the phenomenon with which we are concerned is a kind of pre-sensory[1] experience which only creates the apparatus which later makes qualitative distinctions possible. To avoid the misleading connotations which attach to the term experience it will therefore be expedient to employ a more neutral term to describe the formation of new connexions by the simultaneous occurrence of several afferent impulses. We shall for that purpose adopt as the technical term the word 'linkage'.

5.8. By a *linkage* we shall thus understand the most general lasting effect which groups of stimuli can impress upon the organization of the central nervous system. It implies a physiological effect of external events on that organization, but not necessarily that when these external events occur they already possess any distinct significance for the organism. It is a sort of learning to discriminate which may occur before any discriminations are yet possible, an 'experience' which, though it will later, when the same stimuli occur again, give them special significance for the organism, need at the time as yet have no meaning for the individual.

5.9. When we stress that events producing these linkages need not be in any sense mental or sensory events, we of course do not merely mean that they need not be conscious. It must be remembered that we employ here the term 'mental' in a sense which is wider than, and includes, the conscious (1.67–1.73). The events between which linkages occur need not possess even such a place in the mental order which would make them mental events in this wider sense.

---

[1]This concept of pre-sensory experience must not be confused with the conception of 'pre-sensation' as used by F. R. Bichowski, 1925, and R. B. Cattel, 1930, to describe the 'first conscious effect that can be traced to a stimulus . . . which does not yet possess spatial or temporal quality, that is to say, is not felt to be located in space or time, or to have the definite qualities and relations usually associated with sensations'. (Bichowski, p. 589.)

5.10.   In some respects it might have been preferable instead of introducing the clumsy new term 'linkage', to revive in the same technical sense the old term 'impression'. But this term not only is so much charged with the meaning of a mental experience that it seemed better to avoid it, but it also seemed desirable to choose a term which expressly stressed that all such experience which can give rise to memory must always consist in the creation of connexions between several physiological events. And since all memory consists in the linking of two or more such events it seemed better to describe the effect which produces memory by a term referring to the creation of such links.

5.11.   Although it may sound commonplace that all experience, in the widest sense of the term, causes, and that all memory is based on, the creation of connexions between physiological events representing stimuli, this still requires emphasis, since there exists another view which, just because it is rarely explicitly stated, yet exercises considerable influence and is one of the main supports of the idea of a special mental substance. This view is what might be called the 'storage' theory of memory, the conception that with every experience some new mental entity representing sensations or images enters the mind or the brain and is there retained until it is returned at the appropriate moment.

5.12.   This conception is, of course, part and parcel of the theory of the absolute character of sensory qualities, and connected with the erroneous interpretation of the theory of the specific energy of the nerves, according to which the nature of the process in the different fibres determines the quality of the resulting sensation. Against it we should remember that we know of no physiological mechanism which can retain anything except connexions between different events, and that, therefore, any theory of mind which is to be expressed in physiological terms must use 'experience' and 'memory' in the sense which we stress by employing the term 'linkage.'

5.13.   The theory here developed then assumes that every sensory quality which occurs presupposes the previous occurrence of linkages between impulses which may not yet have been classified as belonging to a particular qualitative group. Even after relatively simple systems of connexions, effecting some measure of classification, have been formed, this system will be constantly modified by new linkages. But as the existing system of connexions becomes more and more complex and more firmly embedded, any new linkage will be less likely to alter its general character.

5.14. One important consequence of this relation between physiological linkages and sensory experience is that there will be implicit in all sensory experience certain relations determined by earlier linkages (i.e., by the influence of the external world on the organism) which have never been the object of sensory experience in the ordinary meaning of this term; and that the order of sensory qualities will be subject to continuous modification by new linkages between impulses which may not lead to sense experience. The epistemological significance of this fact will be examined in the last chapter (8.1–8.27).

5.15. The terminological point discussed in this section has some bearing on the question whether our theory of the determination of sensory qualities can be properly described as 'empiricist'. It seems that in the dispute between the 'empiricists' and the 'nativists' there were really two different issues involved. The first is whether, so far as the individual is concerned, the order of sensory qualities is congenital or acquired by individual experience. On this probably no general answer is possible. The second is, whether the whole sensory order can be conceived as having been built up by the experience of the race or the individual, i.e. whether it is based on the retention of connexions between effects exercised upon them by the external world. With regard to this second question our answer is definitely empiricist (2.16).

5.16. It might at first seem as if this empiricist character of our theory would stand in irreconcilable contrast to the strongly anti-empiricist attitude of the gestalt school with whose arguments our theory is in other respects in close agreement. I am not certain, however, that the opposition of the gestalt school to an empiricist explanation of gestalt qualities as being 'built up' by experience from sensory 'elements' need apply to a theory which, as the theory developed here, traces *all* sensory qualities, 'elementary' as well as gestalt qualities, to the pre-sensory formation of a network of connexions based on linkages between non-mental elements.

2. THE GRADUAL FORMATION OF A 'MAP' REPRODUCING RELATIONS BETWEEN CLASSES OF EVENTS IN THE ENVIRONMENT

5.17. The connexions formed by the linkages between different impulses will evidently reproduce certain regularities in the occurrence of the external stimuli acting on the organism. The network of these connexions will reproduce not any attributes of the individual stimuli (whose identity is determined solely by their capacity of setting up impulses in a particular sensory fibre, or group of fibres), but a sort of record of past associations of any particular stimulus with other stimuli which have acted upon the organism at the same time. While such a record, dependent upon the frequency which in the course of the development of an individual (or possibly species) certain stimuli have occurred together, will reproduce certain relationships between these stimuli determined by the physical differences between them, it will clearly not give a full or correct reproduction of all the relations which can be said 'objectively' to exist between these stimuli.

5.18. We have seen in the first Chapter (1.14–1.19) that a description of the stimuli in physical terms involves a classification of these stimuli based solely on their observed relations towards each other and neglecting any difference or similarity of the response of the organism on which they act. It seems to be in conformity with general scientific procedure to treat only those differences between stimuli which manifest themselves in their relations to other stimuli as differences belonging to the physical world (or as differences constituting the physical order of the universe), and to regard differences and similarities between groups of stimuli which show themselves solely in their effects on certain types of organisms as due to the organization of these organisms. Our present task is to show the kind of classification, or of ordering of the stimuli, which, through the process we have sketched, such an organism is likely to develop.

5.19. The gradual evolution of the mental order involves thus a gradual approximation to the order which in the external world exists between the stimuli evoking the impulses which 'represent' them in the central nervous system. But while conceptual thinking has long been recognized as a process of continuous reorganization of the (supposedly constant) elements of the phenomenal

1

world, a reorganization which makes their arrangement correspond more perfectly with experience, we have been led to the
conclusion that the qualitative elements of which the phenomenal
world is built up, and the whole order of the sensory qualities, are
themselves subject to continuous change. There remains, in consequence, no justification for the sharp distinction between the
direct sensory perception of qualities and the more abstract processes of thought;[1] we shall have to assume that the operations of
both the senses and the intellect are equally based on acts of
classification (or reclassification) performed by the central nervous
system, and that they are both part of the same continuous process by which the microcosm in the brain progressively approxi-
mates to a reproduction of the macrocosm of the external world.

5.20.   The order which the linkages will gradually create in the
central nervous system will, for several reasons, constitute not only
a very imperfect but in some respect even a definitely erroneous
reproduction of the relations which exist between the corresponding physical stimuli. In the first instance, the receptor organs
through which the physical stimuli set up nervous impulses are
imperfectly selective in several respects: the organism possesses
receptor organs which are sensitive for only certain kinds of external events but not for others; and these receptor organs which
it does possess also do not sharply distinguish between stimuli
which are physically different. Physically different events may
stimulate the same receptor organs and set up impulses in the
same sensory fibre, and physical stimuli of the same kind but acting
on different receptors may be recorded as different sensory modalities (1.39). Which external events are recorded at all, and how
they will be recorded, will thus depend on the given structure of
the organism as it has been shaped by the process of evolution.

5.21.   Secondly, the kinds of physical stimuli which will act on a
particular organism, and the relative frequency of the simultaneous occurrence of the different stimuli, will correspond not to
conditions in the world at large, but to conditions in the particular
environment in which the organism has existed. The partial reproduction of the relations between the stimuli acting on the
organism will therefore be a reproduction of those relations only
which appear in a certain sector of the external world, and will
not necessarily be representative of those existing in the whole of it.

[1]Cf., H. Margenau, 1950, p. 54; and H. Werner, 1948, pp. 222–225, 234–236.

5.22. Thirdly, as we have seen already (Chapter IV), one of the most important parts of the 'environment' from which the central nervous system will receive signals producing linkages, will be the *milieu intérieur*, the internal environment or the rest of the organism in which the central nervous system exists. Since the events in the organism will in some degree be co-ordinated with each other and with the events in the external world proper, independently of the functioning of the higher nervous centres, it is inevitable that the relations existing between them should play a large part in shaping the order that will be formed in the higher centres.

5.23. Fourthly, there is no reason to assume that the capacity of the higher centres to form connexions between the neurons in which impulses occur at the same time is uniform throughout those centres. It is probable that the given anatomical structure will facilitate the formation of certain connexions and make the formation of others more difficult (or impossible). The resulting structure of connexions would by this be further distorted or prevented from giving a true reproduction of the relations between even those impulses which unequivocally represent specific physical stimuli.

5.24. Fifthly, as a result of the successive classification of the impulses on several different levels (4.33), the signals reaching the higher and more comprehensive centres will often not represent individual stimuli, but may stand for classes or groups of such stimuli formed at lower levels for particular functional purposes. Any further classification effected at the higher centres will therefore be subject to all the distortions which, for reasons similar to those already mentioned, have occurred on lower levels.

5.25. In discussing the relationships between the network of connexions which will thus be formed, and the structure of external events which it can be said to reproduce, it will be useful sometimes to employ the simile of the *map* which in a somewhat analogous manner reproduces some of the relations which exist in certain parts of the physical world. The picture of the geographical map in this connexion comes so readily to one's mind[1] because of its similarity with the simple arrow diagram which is the most obvious method of schematically depicting the structure of a complex dynamic system whose elements are connected as cause and effect.

[1]For similar uses of the concept of the map see, e.g., E. D. Adrian, 1947, pp. 16–18, and E. C. Tolman, 1948.

5.26. This 'map' of the relationships between various kinds of events in the external world, which the linkages will gradually produce in the higher nervous centres, will not only be a very imperfect map, but also a map which is subject to continuous although very gradual change. It will not only give merely some of the relations existing in the external world, and give in addition some which are different from those which exist objectively, but it will also not give a constant but a variable picture of the structures which it reproduces.

5.27. The different maps which will thus be formed in different brains will be determined by factors which are sufficiently similar to make those maps also similar to each other. But they will not be identical. Complete identity of the maps would presuppose not only an identical history of the different individuals but also complete identity of their anatomical structure. The mere fact that for each individual the map will be subject to constant changes practically precludes the possibility that at any moment the maps of two individuals should be completely identical.

5.28. The conception of 'similarity' between several different systems of relationships which are not completely identical, such as would exist between two maps of this kind, and still more so the conception of varying degrees of similarity, and that of similar positions in similar systems of relationships, present considerable conceptual difficulties. It is the same difficulty which we encounter when we consider the degrees of similarities between various qualities or gestalts. The simile of the map, however, will show what is meant: we can recognize without great difficulty not only the similarity of different maps of the same region, although they may be drawn in different projections, contain different details and refer to different dates, but we shall in general also be able to identify corresponding points on two such maps as referring to the same point in the real world. Two persons discussing the same walk, with different maps of the region before them, will in general encounter no difficulty in understanding each other, although particular points on their route may have different significance for them.

5.29. In the 'map' with which we are concerned, the relevant relations between the individual points are not their spatial relations, however, but solely the paths through which impulses can be transmitted. It is a topological and not a topographical map. It will resemble rather one of those schematic railway maps in which connexions are indicated by straight lines without representing accurate distances. It will resemble a topographical map only in the sense that it will also show where any given movement will lead us.

5.30. For a description of the process by which linkages will gradually produce a map of the relations between the stimuli acting on the organism, the simile of the map, however, soon becomes inadequate, because the classification with which we are concerned will, as we have seen, occur on many successive levels. We should have to think of the whole system of connexions as consisting of many vertically superimposed sub-systems which in some respects may operate independently of each other. Every sub-system of this kind will constitute a partial map of the environment, and the maps formed at the lower levels will serve for the guidance of merely a limited range of responses, and at the same time act as filters or preselectors for the impulses sent on to the higher centres, for which, in turn, the maps of the lower levels constitute a part of the environment.

5.31. While the full and detailed classification of sensory impulses, corresponding to the order of sensory qualities which we know from conscious experience, is effected mainly at the highest centres, we must assume a more limited classification on somewhat similar principles to take place already at the lower levels, where certainly no conscious experience is associated with it. The qualitative order which is familiar to us in its most developed form from our conscious experience will exist in a more rudimentary form on lower levels where we have no direct knowledge of it, but can only attempt to reconstruct it as part of our endeavour to understand the whole hierarchy of the apparatus of classification which culminates in conscious mind. There can be little doubt that we must assume the existence on lower levels of such an order of the sensory impulses somewhat analogous to that revealed by our conscious experience, an order which we can ascertain only from the character of the discriminatory responses of which we are not conscious.

5.32. We have already discussed the significance of this hierarchical order of the central nervous system and the significance which the classification of impulses will have for the operation of the whole (4.11–4.26), and we shall leave aside until the next chapter the question of the 'conscious' character of some of the processes which take place at the highest levels. It will also be seen now, although a fuller discussion of this must wait until later in this chapter, that the difference between what are commonly regarded as merely 'mechanical' and as mental processes respectively is not one of kind but merely one of degree; and that the extent to which a process partakes of the character of the mental will depend on the complexity of the ordering processes which intervene between the stimulus and the response; or rather, since the stimulus-response terminology becomes somewhat inappropriate at this stage, between the excitatory state of the sensory apparatus and the resulting behaviour.

3. THE 'MAP' AND THE 'MODEL'

5.33. We must now consider the manner in which, within a given structure of connexions, the many impulses proceeding at any one moment, can mutually influence each other. Up to this point we have examined only the mutual effects which new afferent impulses arriving more or less at the same time will exercise upon each other. The centres at which such impulses arrive will, however, never be found in an inactive state. As we ascend to higher and higher levels, the function of new impulses arriving there will be less and less to evoke specific responses but increasingly merely to modify and control behaviour in the light of the whole situation, represented not only by simultaneously arriving other impulses but also by the retained picture of the environment. This involves that a sort of record of recently received stimuli is kept at these higher centres.

5.34. As any afferent impulse is passed on to higher levels, it will send out more and more branches which will potentially be capable of reinforcing or inhibiting an ever-increasing range of other impulses. This increasing ramification of every chain of impulses, as it ascends through successive relays to higher levels, will mean that at any moment the general excitatory state of the whole nervous system will depend less and less on the new stimuli currently received, and more and more on the continued course of chains of impulses set up by stimuli which were received during some period of the past. In consequence, an ever-increasing part of the forces determining the response will consist of the pre-existing distribution of impulses throughout the whole system of connected

fibres, while the newly arriving impulses will play a correspondingly smaller part.

5.35. It is a corollary of this steadily increasing influence of the
pre-existing excitatory state that the main significance of any new
stimulus will be that it will alter the general disposition for responding in particular ways to further stimuli, and that less and
less of its effect will consist in producing a specific response. In
other words, a greater and greater part of the effects of impulses
set up by any new stimuli will go to create a 'set' controlling future
responses, and a smaller part directly to influence current
responses. As we reach higher levels, the classification of the
impulses becomes thus less specific to a particular function, and
more general in the sense that it will help to create a disposition to
a certain range of responses to an ever-growing variety of stimuli.

5.36. As the classification becomes thus more 'general' and less
'specific', the classifying event also becomes more and more
definitely a central process while the relations to any particular
peripheral response become at the same time more remote and
round-about. As higher and more embracing centres are reached
the effect of any newly arriving afferent impulse on the central
process will become more and more important compared with its
direct effect on peripheral responses. We must think increasingly
in terms of a continuous central process which at any moment will
merely be somewhat modified by the newly arriving afferent impulses and only part of which will, as it were, continuously spill
over into efferent signals producing peripheral responses.

5.37. It thus will be the totality of all the different impulses
proceeding at any given moment in the higher centres which
determines what is to be the response to any newly arriving
impulse. Since all these impulses thus act as a sort of representation
or picture of the momentary environment, to which the response
to any new impulse is adjusted, it is not fanciful to describe the
whole as an apparatus of orientation.[1] By providing a reproduction of the environment in which the organism moves at the
moment, it adjusts the responses to those elements in the environment which are represented in it.

5.38. It seems probable that at these higher centres some of the
impulses representing external stimuli continue for a time to circu-
late in some manner through the same fibres after the stimulus has

[1] H. Kleint, 1940, p. 40.

ceased to operate, and to indicate the presence of an external object although it no longer acts as a stimulus. This may be brought about by the impulses which represent the total sensory situation of a moment becoming associated with each other and mutually evoking each other, until this representation of a given object is wiped out by some new stimulus indicating that a different object now occupies the same point in the spatial order. (See 3.15 and 3.34 above.)[1]

5.39. It is indeed a difficult problem why and in what circumstances a given set of representative impulses will ever lead to the expectation of a more or less constant environment, or produce the persistence of a given picture of the particular environment in which the organism exists. The explanation probably is that, as suggested, certain constellations of impulses mutually support each other, or that by a sort of circular process they will tend to re-evoke themselves rather than a different constellation corresponding to a different environment. The interaction of the chains of associations attached to the different impulses and groups of impulses will effect some kind of selection from the infinite range of possibilities which the several elements of the complex picture would tend to evoke.

5.40. Whether this actually is the case or not, we must certainly distinguish between two different kinds of physiological 'memory' or traces left behind by the action of any stimulus: one is the semi-permanent change in the structure of connexions or paths which we have already discussed and which determines the courses through which any chain of impulses can run; the other is the pattern of active impulses proceeding at any moment as a result of the stimuli received in the recent past, and perhaps also merely as part of a continuous flow of impulses of central origin which never altogether ceases even if no external stimuli are received.

5.41. The pattern of impulses which is traced at any moment within the given network of semi-permanent channels may be regarded as a kind of model of the particular environment in which the organism finds itself at the moment and which will

[1]This important point can here only just be touched upon since a fuller discussion would require a more detailed consideration of physiological detail than would be appropriate here. For an important attempt at elaboration in this direction see now D. O. Hebb, 1949.

enable it to take account of that environment in all its movements. This 'model', which is formed at any moment by the active impulses, must not be confused with what we have called the 'map', the semi-permanent connexions representing not the environment of the moment but the kind of events which the organism has met during its whole past. This distinction between what for lack of better terms we shall continue to describe as the 'map' and the 'model' respectively, both of which are reproductions of relations between events in the outside world, is so important that it requires some further elucidation.

5.42. The semi-permanent map, which is formed by connexions capable of transmitting impulses from neuron to neuron, is merely an apparatus for classification or orientation, capable of being called into operation by any new impulse, but existing independently of the particular impulses proceeding in it at a given moment.[1] It represents the kind of world in which the organism has existed in the past, or the different *kinds* of stimuli which have acquired significance for it, but it provides by itself no information about the particular environment in which the organism is placed at the moment. It is the apparatus of classification in what we have called its static aspect (2.21) and would continue to exist (if this were possible) if at any moment the central nervous system were completely at rest.

5.43. This semi-permanent structure provides the framework within which (or the categories in terms of which) the impulses proceeding at any time are evaluated. It determines what further impulses any given constellation of impulses will set up, and represents the kinds of classes or 'qualities' which the system can record, but not what particular events will be recorded at any moment. This structure itself in turn is liable to change as a result of the impulses proceeding in it, but relatively to the constantly changing pattern of impulses it can be regarded as semi-permanent.

5.44. Within this structural framework of paths the flow of impulses will at any moment trace a further pattern which will

[1]This manner of stating the difference is correct only on the assumption that the connexions involve structural changes and are not merely 'functional', i.e. produced by something like a continuous circuit of impulses (2.47). If the connexions should prove to be based not on structural changes but on some functional change of the latter kind, this would probably not affect the principle of the distinction made above but would make the description of the mechanism much more difficult.

have significance only by its position in that structural framework within which it moves. This 'model' formed by the moving impulses or by a particular operation of the apparatus of classification manifests the latter in its dynamic aspect. Its nature is, of course, limited by the possibilities which the structural 'map' provides, by the connexions or channels which exist; but within these limits its character will be determined by the combined effects of the active impulses.

5.45. The response to any newly arriving sensory impulses will thus depend not only on the semi-permanent map formed by the network of connexions; it will also depend on the pattern of the impulses already proceeding at that movement within the pattern of channels; and it will be the position of the former within the latter pattern which will determine the significance of the new impulses. The complete apparatus of orientation thus consists of a structure of which a certain part will be activated, or of a sort of model within a model which has significance only by its position within that model, and which adjusts the responses to any new stimulus not only to the general significance which stimuli of that sort will possess in any circumstances, but to the particular significance which they will possess in the situation existing at the moment.

5.46. This relationship between our 'model' and our 'map' is the same as that which, in a game of war (*Kriegspiel*) played with symbolic figures on a map, exists between the patterns traced on the map by the figures and the map itself. Or we may think of a game of Nine Men's Morris where similarly the relative position of the men to each other is significant only with respect to the pattern on the board on which the game is played. To make the comparison closer we would have to imagine that the 'men' in either instance are individually undistinguishable from one another apart from their position on the map or the board, and that at any moment, in addition to the men present as the result of the preceding moves, new men may appear at certain points and, finally, that in moving the men leave traces which gradually alter the pattern on the map or the board. The important point, with respect to which these illustrations correspond with our case, is that the pattern traced by the movement of the men is significant *not* by its shape in space but solely by its relation to the other pattern within which those movements take place (cf. 2.5).

5.47. A closer parallel to our case would in some respects be provided by a system of pipes or tubes through which move columns of a pliable substance. If we assume that at many points of interconnexion these tubes are joined by 'afferent' tubes which can bring in from the outside new columns of the substance, and by 'efferent' tubes which may drain such columns from the system, that at any junction the columns may divide, and that the direction of the pressure of the columns meeting at such junctions[1] will decide in which further direction they will jointly move on, we get an approximate representation of the situation with which we have to deal. We might even complete the picture by assuming that, e.g., because the system of tubes is bored into some yielding material, pressure from adjoining tubes may lead to some new channels being formed through which the moving substance first seeps and then moves freely. It will then be the pattern which the moving columns trace within (and relative to) the pattern formed by the system of tubes, which will correspond to the pattern traced by the nervous impulses within (and relative to) the structure of connected fibres.

5.48. The relation which exists between our 'model' and the 'map' may also be compared in some respects to the relation existing between some complex geometrical structure and the system of co-ordinates with reference to which it can be defined. The essential characteristics of the structure will be described in terms of an equation which can be interpreted with reference to many possible systems of co-ordinates, and the actual structure will appear different according as we represent it within different (say Cartesian or polar) co-ordinates. What is significant about the structure of our 'model' is not the actual relations in space between the impulses, but solely their relations to the structure of connexions, relations which correspond to those expressed by the equation by which a given structure is defined in analytical geometry.

5.49. Once such a continuous reproduction of the environment is maintained in the highest centres, it becomes the main function

---

[1]Any mechanical model of this kind is misleading in suggesting a transmission of substance or energy,while in the transmission of nervous impulses we have, of course, to deal with a case of 'trigger action' where the connexions between the neurons merely effect a release of energy stored up in the individual neuron (3.8).

of the sensory impulses to keep this apparatus of orientation up to date and capable of determining the responses to particular stimuli in the light of the whole situation. The classification of these impulses is then no longer specific to particular functions, but has become general in the sense that any one of them may, by its position in the comprehensive pattern, exercise some influence on practically any response. The classified impulses proceeding at any moment operate as symbols representing the significance of the stimuli which have evoked them, for any behaviour which a newly arriving stimulus will tend to evoke, or which would be determined by the joint effect of the multiplicity of processes set in train by stimuli recorded earlier.

4. ASSOCIATIVE PROCESSES

5.50. The pattern of nervous impulses which at any moment will be traced within the structure of connected fibres is, of course, a constantly changing pattern. The representations of the different part of the environment which the impulses produce will derive their significance exclusively from the fact that they tend to evoke certain other impulses. Each impulse representing an event in the environment will be the starting point of many chains of associative processes; in these the various further impulses set up will represent events which in the past have become connected for the individual with those which are represented by the impulses which evoke them. If each of these several chains of associative processes were allowed to run its course, unaffected by other similar chains which have been set up by other impulses (which were either part of the same initial position or which are produced by new stimuli), they would tend to produce representations of a variety of consequences which would follow from the initial environment, rather than present a definite picture of that environment.

5.51. The pattern of impulses formed within the structure of connexions will thus function as an apparatus of orientation by representing both the actual state of the environment and the changes to be expected in that environment. This, of course, must not be understood to mean that representations of several different states of affairs can exist simultaneously. It means that each part and the whole of the representation of the existing environment derive their significance from the penumbra of possible consequences attaching to them: what gives every element or group of elements of which the total situation exists their sensory value is their following representing their various potential effects.

5.52. It is particularly important in this connexion not to fall back upon the traditional conception of the individual impulses as such corresponding to particular mental qualities or to conceive of associative processes as simple chains of impulses where physiological elements correspond to mental units. The physiological impulse owes its mental quality to its capacity of evoking other impulses, and what produces the succession of different mental qualities is the same kind of process as that which determines the position of the impulses in the order of mental qualities: it possesses such a quality only because it can evoke a great variety of associated impulses. Association, in other words, is not something additional to the appearance of mental qualities, nor something which acts upon given qualities; it is rather the factor which determines the qualities.

5.53. The mental qualities which succeed each other in the course of associative processes do therefore not correspond to the units between which physiological connexions exist. The sequence of individual mental images (or reproductions) is rather the resultant of the interaction of a multiplicity of streams of impulses, and every new mental quality which is thus evoked will be the effect of not only of those physiological impulses preceding it which have themselves been fully evalued, but also of those which have merely contributed to the evaluation of the former, and of others which have not received sufficient support to obtain a distinct following. Even the simplest succession of mental qualities which appear directly to evoke each other must probably be conceived as the resultant of a complex process of convergence of many impulses.

5.54. We must probably assume that in order than an impulse should be able to produce its own distinctive following and thus obtain a distinct place in the qualitative order it will as a rule require the support from other impulses whose followings closely overlap with its own. But while those to which these applies will occupy, as it were, the centre of the stage, those which are not sufficiently supported to produce their complete following will nevertheless exercise some influence on the further course of the associative processes. Even though only the concentrated stream of impulses which forms the 'foreground' will be fully evaluated, its effects will also depend on the less distinct background.

5.55. So far as the appearance in consciousness of successive images is concerned, of which we think in the first instance when we mention 'associations', this means, of course, that what further images will be evoked by a conscious event will depend not only on it but also on much which is not conscious. But, as we shall see more fully in the next chapter, the difference between conscious and non-conscious events is a difference of the same kind as that which appears on the highest level as the difference between more or less fully evaluated events. In all these instances the effect which on a given level will be produced by an event occupying a distinct position in the order of that level will depend not only on itself but also on the substructure of less fully discriminated events on which it rests and which forms its background.

5.56. The different associations attaching to individual impulses, and still more those attaching to any one of a group of simultaneous impulses, will often not only not be convergent but even conflicting; and not all the representations which will form part of the following of the elements of the complete situation will be capable of simultaneous realization, or would produce a significant new pattern if they did. Since from each element of the structure of connected fibres impulses can pass in a great variety of directions, the initial stream of impulses would merely diffuse and dissipate itself if the overlapping of the following of the many different impulses did not determine a selection of some among the many potential paths on which they might travel.

5.57. This selection will be brought about by the fact that, where the followings of the representations of the different parts of the environment overlap, the corresponding impulses will reinforce each other by summation (3.13) and by their joint effects evoke sequences of representations which otherwise would have remained merely 'potential'; while in so far as the various elements set up divergent or even conflicting (mutually inhibiting) tendencies, these flows of impulses will mutually neutralize each other.

5.58. The representation or model of the environment will thus constantly tend to run ahead of the actual situation. This representation of the possible results following from the existing position will, of course, be constantly checked and corrected by the newly arriving sensory signals which record the actual developments in the environment. The newly arriving impulses, on the other hand, in turn will always be evalued against the background of the expectations set up by the previously existing pattern of impulses.

5.59. The representations of the external environment which will guide behaviour will thus be not only representations of the actually existing environment; but also representations of the changes to be expected in that environment. We must therefore conceive of the model as constantly trying out possible developments and determining action in the light of the consequences which from the representations of such actions would appear to follow from it.

5.60. We shall have further to consider the character of these associative processes in the next chapter when we consider conscious thought, and again in the last chapter when we shall have to examine the nature of explanation. At this stage our concern is merely to emphasize that processes essentially analogous to the processes of association which are familiar to us from conscious thought, must be assumed to play a similar rôle already on pre-conscious levels. The processes of classification with which we are concerned, and which will determine conscious as well as unconscious responses, constitute classifications of complex situations by the joint results to be expected from the simultaneous occurrence of the elements; and this in turn involves the representation of the range of expected results by a pattern of impulses essentially in the same manner in which the actual environment is represented by such a pattern.

5.61. The representation of the existing situation in fact cannot be separated from, and has no significance apart from, the representation of the consequences to which it is likely to lead. Even on a pre-conscious level the organism must live as much in a world of expectation as in a world of 'fact', and most responses to a given stimulus are probably determined only *via* fairly complex processes of 'trying out' on the model the effects to be expected from alternative courses of action. The reaction to a stimulus thus frequently implies an anticipation of the consequences to be expected from it.[1]

5.62. It is these chains of symbolic representations of the consequences to be expected from a given representation of events which we must conceive as constituting those 'symbolic processes

[1] Cf., R. Dodge, 1933, V. von Weizsaecker, 1947, p. 136.

in the brain' which physiological psychology has been led to postulate[1] in order to account for the complex adaptive responses, and to explain the delays involved between stimulus and response even on levels where there is no ground for assuming the presence of consciousness, or where we know that the responses take place without our being aware of the stimulus which has evoked them.

6. MECHANICAL AND PURPOSIVE BEHAVIOUR

5.63. The principles by which the transmission of the individual impulses in the central nervous system is determined are of a kind which might well be described as 'mechanical' in the most general meaning of the word; yet the result of the interplay of these transmissions in the integrated nervous system will clearly show characteristics which are not only very different from, but in some respects even the opposite of, those which we commonly associate with that term. It will therefore be useful explicitly to enumerate the points on which the multiple process of classification will act in a manner different from what we ordinarily regard as mechanical.

5.64. By a 'mechanism' or a 'mechanical process' we usually understand a complex of moving parts possessing a constant structure which uniquely determines its operations, so that it will always respond in the same manner to a given external influence, repeat under the same external conditions the same movements, and which is capable only of a limited number of operations. Such a mechanism cannot 'purposively' adapt its operations to produce different results in the same external conditions; and it is essentially 'passive', in the sense that which of the different operations of which it is capable it will perform will depend exclusively on the external circumstances.

5.65. In all these respects the operation of a system organized on the principles here described will show opposite characteristics. It will, as a result of its own operations, continuously change its structure and alter the range of operation of which it is capable. It will scarcely ever respond twice in exactly the same manner to the same external conditions. And it will as a result of 'experience' acquire the capacity of performing entirely new actions. Its actions will appear self-adaptive and purposive, and it will in general be 'active' in the sense that what at any given moment will

[1]Cf., C. T. Morgan, 1943, p. 112.

determine the character of its operation will be the pre-existing state of its internal processes as much as the external influences acting on it.

5.66. Since the structure of connexions in the nervous system is modified by every new action exercised upon it by the external world, and since the stimuli acting on it do not operate by themselves but always in conjunction with the processes called forth by the pre-existing excitatory state, it is obvious that the response to a given combination of stimuli on two different occasions is not likely to be exactly the same. Because it is the whole history of the organism which will determine its action, new factors will contribute to this determination on the later occasion which were not present on the first. We shall find not only that the same set of external stimuli will not always produce the same responses, but also that altogether new responses will occur, if we regard as one response not the movement of an individual muscle but the whole complex of co-ordinated movement of the organism.

5.67. The appearance of such new forms of behaviour is the effect of the circumstance already noted (4.25, 4.49) that the individual motor impulses sent out by the higher centres will be signals not for particular movements but for classes or kinds of actions, and that the particular movement that will occur will be determined by the joint effect of many such general 'directives'. The signal for a particular succession of movements of various muscles may for instance be modified by other signals indicating that it is to take place fast and with the avoidance of noise, or that other types of modification of the basic pattern are required. Any one particular movement will thus be determined by the higher centres signalling the different 'qualities' required from the action, and these different 'qualities' of behaviour will be closely interwoven (4.46) with the qualities ascribed to events in the external world.

5.68. The adaptive and purposive behaviour of the organism is accounted for by the existence of the 'model' of the environment formed by the pattern of impulses in the nervous system. In so far as this model represents situations which might come about as the result of the existing external situation, this means that behaviour will be guided by representations of the consequences to be expected from different kinds of behaviour. If the model can preform or predict the effects of different courses of action, and pre-select among the effects of alternative courses those which in the existing state of the organism are 'desirable', there is no reason why it should not also be capable of directing the organism towards the particular course of action which has thus been 'mapped out' for it.[1]

5.69. In order that the 'desired result' should operate as a cause determining behaviour, it must be evoked by, or form part of the following of, the reproduction of the actual environment and of the governing state of drive or urge. It must be a representation of the innumerable combinations of possible outcomes of the existing situation which the convergent associations tend to evoke —associations which are attached to the elements of the representation of that environment, and which give these representations their significance or meaning. The 'desired' result will be singled out from the many possible outcomes of the existing situation because it will form part of the following not only of the environment but also of the 'urge' or 'drive' for a certain class of results. That representation of the results which seem both possible in the existing external situation and 'desirable' in view of the state of the organism will thus be strengthened by the summation of two different streams of nervous impulses. This result will be represented with greater strength and distinctness and will thereby come to exercise the determining influence on further behaviour.

5.70. Such a representation of certain possible outcomes of the existing situation, which are strengthened because they appear desirable, constitute, however, only a first step towards purposive behaviour. In most situations there will exist many possible courses of action which, in the sense that they have in the past become associated with the achievement of that goal, will appear to be 'directed' towards a desirable goal; but only some of these courses of action will be appropriate in the particular situation. There will, in general, also exist more than one possible goal of the desired kind, and more than one way of achieving any one goal. The determination of purposive action involves, therefore, a further process of selection among the various different courses of action which might satisfy the initiating urge.

5.71. The interpretation of the pattern of moving impulses as a model of the environment, which can try out possible developments in that environment, suggests answers to both these problems. We shall first consider the mechanism by which will be

[1]On this and the following see K. J. W. Craik, 1934, pp. 51 ff.

eliminated inappropriate moves which, though they would produce the desired result if those features of the environment which
evoke that response were alone present, yet cannot in fact produce
that result because other elements in the environment constitute
an obstacle to the completion of this course of action.

5.72. Such a situation would have its neural counterpart in the
impulses representing different parts of the environment tending
to set up contradictory or incompatible chains of associations
which will mutually blot each other out. Because representations
of possible developments in the environment will have significance
or meaning only as parts of an ordered picture of that environment, the various chains of associations set up by the elements of
the representation of a complex situation will produce meaningful
results only if they fit into the general order of such representations.
The general spatial and temporal order of that representation
will, e.g., require that either one thing or another must be in a
given place, or that a thing must be either at rest or in motion,
etc., etc. In so far as the chains of associations, set up by different
parts of the representation of the environment, lead in this sense
to conflicting results, e.g., to the representation of two different
things as being in the same place at the same moment, or of two
incompatible attributes being attached to a particular thing,
these tendencies will neutralize each other: no distinct representation will result which could become the starting point for further
associations.[1]

5.73. Of the many sequences of representations which different
parts of the representation of the existing environment tend to
evoke, only some will thus in fact lead to representations of meaningful results. The mechanism which in this manner eliminates
abortive courses of action must also prevent, however, that at any
moment more than one of the possible alternative courses of
actions should be fully represented, although the model might
successively try out different courses of action. How will it be
determined which of the various courses promising to produce a
desirable result will in fact be selected?

[1]Cf., K. J. W. Craik, 1943, p. 57: 'as a result of such interactive or associative
processes we might have, for example, $A=B$, $B=C$, $A=C$, where A, B, and C
are neural patterns claiming to represent external things or processes. These
patterns clearly cannot remain simultaneously excited; inconsistency means a
clash in the interaction patterns'.

5.74. The first point to be noticed is that the desirability of one particular result will not be the only factor of an affective character which will be operative in such a situation. Most of the different courses of action between which the organism will have to choose, and most of the intermediate stages through which these different courses of action will lead it, will also possess affective qualities— they will themselves be either attractive or repellant in various degrees. In particular, the representation of the effort involved in the different courses of action will normally be charged with the representation of pain, or operate as something to be avoided, unless compensated for by the greater attraction of the result. The interaction of all these forces in the end will bring it about that from the possible courses the 'path of least resistance' will be chosen; while all the unduly painful courses will be avoided which might produce the same result, as well as courses leading to alternative results but requiring greater effort.

5.75. That such guidance by a model which reproduces, and experimentally tries out, the possibilities offered by a given situation, can produce action which is purposive to any desired degree, is shown by the fact that machines could be produced on this principle (and that some, such as the predictor for anti-aircraft guns, or the automatic pilots for aircraft, have actually been produced) which show all the characteristics of purposive behaviour.[1] Such machines, however, are still comparatively primitive and restricted in the range of their operations compared with the central nervous system. They would be able to take account of only a minute fraction of the number of different facts of which the central nervous system takes account, and they would lack the capacity of learning from experience. But although for this reason such machines cannot yet be described as brains,[2] with regard to purposiveness they differ from a brain merely in degree and not in kind.[3]

5.76. It is notoriously difficult to discuss purposive behaviour without employing terms which suggest the presence of conscious mental states. The phenomenon of purposive action, however, does not presuppose the existence of an elaborate mental order

[1] See also W. G. Walter, 1950.
[2] C. Sherrington, 1949.
[3] K. J. W. Craik, 1943, p. 51. See also N. Wiener, 1948a and 1948b and W. S. McCulloch, 1948.

like the one which we know from conscious experience, and still less the presence of consciousness itself. Some degree of purposiveness can be attained by structures infinitely more simple than those which constitute the mental order, and once we have reached the degree of complexity in the ordering of stimuli and responses characteristic of the latter, it does not present a new or separate problem.

7. THE MODEL-OBJECT RELATIONSHIP

5.77. It will be useful further to elucidate the character of the relationship between the 'model' and its object, and to illustrate the possibilities of 'reproducing' certain features of a complex structure within certain parts of the same structure, by constructing an imaginary and greatly simplified model of the model-object relationship itself. To be quite satisfactory for our purposes this super-model should be conceived in strictly physical terms, that is, it should be built up from elements whose properties are all defined in terms of explicit relations to the other elements of the system, and which possessed no phenomenal or sensory properties whatever. But as such a purely abstract model would give little help to the imagination, it will be necessary to resort in some measure to visual imagery.

5.78. While using the conception of a model for this purpose, we must, of course, avoid the suggestion, originally connected with the word model, that it must be the creation of a thinking mind. Since the purpose of introducing the conception is to show how the human mind itself may in a certain sense be conceived as a model of the macrocosm within which it exists, there must be no such three terms as object, model, and the modeller. Our task is rather to show in what sense it is possible that within parts of the macrocosm a microcosm may be formed which reproduces certain aspects of the macrocosm and through this will enable the substructure of which it forms part to behave in a manner which will assist its continued existence.

5.79. We shall conceive for this purpose of a self-contained system or universe consisting of a cloud of particles which individually differ from each other solely by the different effect which in different combinations or constellations they will exercise upon each other. With regard to this universe we shall assume that we possess the capacity of the Laplacean 'demon' of knowing of every particle all its relations to other particles, and therefore of being able to identify each individual particle by the different effects it will have in all conceivable circumstances. Particles of the same kind would mean particles which in all conceivable circumstances could be substituted for one another without thereby altering the course of events.

5.80. Strictly speaking we ought, of course, not even conceive of what we described as a 'cloud' of particles as being arranged in perceptual space, but should describe the relations of these particles to each other in terms of their acting upon each other in particular ways. But since we must resort to representation in perceptual terms, we cannot dispense with the familiar spatial order, and for the purposes of visualizing the order of our universe it will even be useful to imagine that the individual particles possess perceptible identifying marks, such as different colours, indicating the class to which they belong.

5.81. Among the different properties which the different kinds of particles will possess, one of the most important will be their different capacities of combining with other particles of the same or of different kinds into more or less stable structures, which, as structures, will show their own peculiar relations to other particles or structures of particles. By stability of such structures we mean the probability of their persistence in the face of the action upon them by the environment. All the possible structures which groups of particles may form within our universe will in this sense require for their persistence certain environmental conditions, which in the case of some kinds of such structures may almost always be satisfied, while for others they may be of rare occurrence. On the whole, the more complex the structure, the greater will be the number of external influences capable of destroying it, and the more special the circumstances required for its continued existence.

5.82. Some such structures may persist, not because their coherence resists most external circumstances, but because they will move away from such agents before they are destroyed. In the familiar world a drop of mercury is likely to preserve its cohesion because it is apt to 'get out of the way' of any mass which might squash it, and a leaf avoids being torn to shreds by a high wind by taking up a position of least resistance. There is, of course, nothing 'teleological' in these movements which assist the persistence of such structures; but the cause which would destroy them if it acted in its full force or for any length of time at the same time removes them from its influence.

5.83. In general, if the conditions required for the persistence of any more complex structure are not likely always to prevail at the place where it happens to be, it will continue to exist only if (and we shall encounter such substructures, except as temporary phenomena, only in so far as) they can respond appropriately to certain events, and even in some measure anticipate their occurrence, i.e. perform the appropriate response as soon as certain other events occur which indicate the imminence of the harmful one. It is conceivable, though not very likely, that such structures may persist because they just happen thus to respond to all or most of the events which usually precede those which would destroy them.

5.84. The chance of persistence of any given structure will evidently be increased if it not only happens to respond appropriately to harmful or beneficial influences and to some symptoms of such factors, but if it also possesses the capacity of retaining a 'memory' of the connexions between events which frequently precede such influences and these influences themselves, and thereby becomes capable of 'learning' to perform the appropriate response whenever those signals appear. Relative complex structures which, without this capacity could not exist long, may through it acquire a considerable degree of stability.

5.85. Correct anticipation of future events in the environment can rarely be based on a single present event but will as a rule have to take into account a combination of many present events. It involves thus different responses not only to different individual events but also to different combinations of such events according as they are likely to produce one result or another. Any mechanism which makes the structure respond to different combinations of external events according to the different further events they are likely to produce, implies that there exists inside the structure a system of relationships between events caused by the external circumstances which is in some measure structurally equivalent to the system of relationships which exists between those external events. Such an internal structure which reproduces some of the relations between the outside events we have called a model.

5.86. It is conceivable that a structure endowed with the capacity of retaining experienced connexions might learn separately the appropriate responses to most of the possible combinations of events. But if it had to cope with the complexity of its environment solely by classifying individual events and learning separately for every combination of such events how to respond, both the complexity of the model required and the time needed for building it up would be so great that the extent to which any given structure could learn to adapt itself to varying circumstances would be very limited.

5.87. It is in this connexion that the various processes of multiple classification which we have described, and the phenomena of 'transfer' and 'generalization' which they make possible, greatly extend the predicting capacity of any models that can be formed from a limited number of elements. Whenever the classifying mechanism treats as alike, or as alike in certain circumstances, any group of events, it will be able to transfer any experience with any one of them to all of them. The process of learning is thereby greatly abbreviated and the complexity of the apparatus required to cope with a given variety of situations is greatly reduced.

5.88. If, for instance, the combinations of any one of the events $(a_1, a_2, a_3, \ldots a_k)$ with any one of the group of events $(b_1, b_2, b_3, \ldots b_m)$ and any one of the group of events $(c_1, c_2, c_3, \ldots c_n)$ produce X, there will be $k \cdot m \cdot n$ (or if $k = m = n = 10$, one thousand) possible different combinations of events for each of which it would have to be learnt separately that they produce X. But once it has already been learnt that in all other respects all members of class A (i.e., $a_1, a_2, a_3, \ldots a_k$) are equivalent, and the symbol o substituted for them, and similarly for all members of the class B the symbol p, and for all members of the class C the symbol q, then the experience that $o + p + q$ produces X will be sufficient to predict the outcome of the $m.n.k$ different combinations of individual events.

5.89. It is thus the process of multiple classification which builds the model. What we have before called the 'map', the semi-permanent apparatus of classification, provides the different generic elements from which the models of particular situations are built. The term 'map', which suggests a sort of schematic picture of the environment is thus really somewhat misleading. What the apparatus of classification provides is more a sort of inventory of the kind of things of which the world is built up, a theory of how the world works rather than a picture of it. It would be better described as a construction set which supplies the parts from which the models of particular situations can be built.

5.90. The model-building by such an apparatus of classification simplifies the task and extends the scope of successful adaptation in two ways: it selects some elements from a complex environment as relevant for the prediction of events which are important for the persistence of the structure, and it treats them as instances of classes of events. But while in this way a model building apparatus (and particularly one which can be constantly improved by learning) is of much greater efficiency than could be any more mechanical apparatus which contained, as it were, a few fixed models of typical situations, there will clearly still exist definite limits to the extent to which such a microcosm can contain an adequate reproduction of the significant factors in the macrocosm.

5.91. It is, from this angle, no more than a fortunate accident that the different events in the macrocosm are not all fully interdependent to any significant degree, but that as a rule it is possible to base predictions of certain kinds of events on a mere selection from the totality of all events. If it were not possible for practical purposes to isolate quasi-self-contained substructures, containing no more parts which significantly affect the relevant result than can be reproduced, or matched point by point, by 'representative' elements within our organism, prediction and purposive adaptation would be impossible. But while it seems that the complexity of the relations which must be taken into account for most purposes are sufficiently limited to make it possible for some structures to 'contain' adequate reproductions of them, this can evidently not be universally true. We shall have to concern ourselves in the last chapter with the significance of the fact that any coherent structure of this kind which within itself contains a model guiding its actions, must be of a degree of complexity greater than that of any model that it can contain, and therefore than that of any object it can reproduce.

CHAPTER VI

CONSCIOUSNESS AND CONCEPTUAL THOUGHT

I. CONSCIOUS AND UNCONSCIOUS MENTAL PROCESSES

6.1. We have used the term 'mental' to describe all processes which involve a classification of events in terms of a qualitative order, similar to that which we know from our subjective sensory experience, and differing from the physical order of these events. The sphere of 'mental phenomena' in this sense is far more extensive than that of conscious phenomena and includes many events which are undoubtedly not conscious. This definition of the mental raises the problem of the determination of this mental order in a form in which it would also arise from a consistently pursued behaviourist approach. We know that an order which is at least similar to that which we know from our conscious experience manifests itself not only in the behaviour of lower animals, with regard to whom we have no ground for assuming that the events so ordered are accompanied by anything which could be described as conscious experience, but also in many responses of our own body where we know that the initiating stimuli do not give rise to any conscious experience.

6.2. Up to this point we have only incidentally referred to the conscious character which distinguishes some of those mental events, and have concentrated on the general character of the qualitative order of all mental events, whether they are conscious or not. What has already been said on this point should justify such a use of the term 'mental' as including both unconscious and conscious events, even though in the past such a use of the term has been often explicitly condemned by psychologists (1.73). We must now, however, attempt at least some indication of the additional characteristics or attributes which distinguish conscious from unconscious mental events.

132

6.3.   While all qualitative discriminations thus imply the presence of a mental order in the sense in which we have used this term, it does not necessarily involve that the individual is 'conscious' or 'aware' of these processes. Consciousness, in the sense in which this term is synonymous with awareness,[1] is an attribute which attaches only to some but not to all mental events. But although we all know what we mean when we say that we are 'conscious' or 'aware' of certain experiences, it is exceedingly difficult to state precisely of what the peculiar attribute of such conscious mental events consists.

6.4.   It may be that it is impossible to give a satisfactory definition of what consciousness 'is', or rather that this is a phantom-problem of the same kind as the 'problem' of the 'absolute' character of the sensory qualities. We shall endeavour to avoid this difficulty by not asking what consciousness 'is' but by merely inquiring what consciousness does. In other words, we shall be concerned solely with the differences which exist between behaviour which we know to result from conscious mental processes and behaviour produced by unconscious mental processes. Before examining this question in any detail, three propositions may be stated which will probably command fairly general assent.

6.5.   It seems clear, in the first instance, that, although the distinction between conscious and non-conscious processes refers originally to different kinds of processes occurring in ourselves, we also employ it to describe the differences which we observe in the behaviour of other people. We know from our subjective experience that there are differences between those actions of our body which we 'deliberately' control and those which take place without our control, and we use this knowledge to distinguish similarly between correspondingly different actions of other people. Although we possess no certain criterion on which we can base this distinction, we are sufficiently familiar with the two kinds

[1] J. G. Miller, 1942, p. 43, treats 'awareness of discrimination' as a definition of consciousness (or rather, unawareness of discrimination as a definition of unconsciousness). It is the last of sixteen definitions of consciousness which he lists, and he describes it as 'the golden meaning of the word for those who admit the validity of introspective testimony,' and later (p. 294) as 'the fundamental meaning of conscious' to most behaviourists. But while this distinguishes this meaning from others, it hardly defines the concept, since 'aware' seems merely a synonym for 'conscious' in this sense.

of actions to be able to attribute consciousness with fair assurance to persons whom we observe acting in certain ways.

6.6. There can, secondly, be little question, although in extreme instances the difference between conscious and unconscious mental events appears to be so complete as to make the difference appear to be one of kind, that there exist many forms intermediate between fully conscious and fully unconscious mental events which make this difference one of degree. Consciousness is evidently capable of many different degrees of intensity, and between the clearly conscious and the clearly unconscious states there exist many forms of semiconscious events with regard to which it is difficult to decide whether they ought to be described as conscious or not.[1]

6.7. Thirdly, it will probably be readily conceded that all conscious events possess in the highest degree the attributes characteristic of all mental processes: conscious responses are to an especially high degree modifiable and purposive, different conscious events are very closely related to each other and very fully discriminated from each other; furthermore, this discrimination is even less 'specific' and more 'general', in the sense in which we have used these terms (5.36, 5.49) than is true of other mental phenomena; and they are even more likely to produce completely new complexes of behaviour than is true of unconscious mental processes.

2. CRITERIA OF CONSCIOUSNESS

6.8. What, then, are those special attributes of conscious behaviour by which we distinguish it from behaviour which also appears to be co-ordinated and purposive but of which the acting person is not 'aware'? Such unconscious behaviour may occur either because the person's attention is at the particular moment otherwise engaged, or because he is altogether unconscious as is the case in some somnambulic states and hypnotic trances. There appear to exist three *prima facie* differences between such unconscious and conscious behaviour which we may provisionally describe by saying that in conscious behaviour a person will, (*a*) be able to 'give an account' of what he is or has been doing, (*b*) be able 'to take account' in his actions of other simultaneous

[1] J. G. Miller, 1942, p. 166.

experiences of which he is also conscious, and, (c) be guided to a large extent not only by his current perceptions but also by images and reproduction of circumstances which might be evoked by the existing situation.

6.9.   When we say that a person is able to 'give an account' of his mental processes we mean by this that he is able to communicate them to other people by means of 'symbols', that is by actions which, when perceived by other people, will occupy in their mental order a position analogous to that which they occupy in his own; and which, in consequence, will have for those other persons a meaning similar to that which it possesses for him.

6.10.   The possibility of such communication between different persons is not only indicative of the presence of consciousness but the symbolisms employed for that purpose may also be an important factor which helps to raise the discrimination to that higher degree of clarity and precision which distinguishes conscious experience. The connexions between sensory impulses and the highly developed apparatus of expression which man possesses undoubtedly greatly extend the means of classification available to him, and they are probably of the greatest importance in making abstract thought possible. They are also especially important because, in learning the system of symbols developed by his race, the individual can utilize, in ordering his current experience, not only his own experience but in some measure also the experience of his race.

6.11.   But although communication (or at least communication by language proper, as distinguished from communication by gestures, facial expression, etc.) will as a rule be the result of conscious processes and 'give an account' of such processes, this means no more than that communication, being normally itself conscious action, is connected with (or can be influenced by) all other conscious processes. As a criterion of consciousness the possibility of 'giving an account' of conscious processes is therefore no more than a special instance of the second of the three criteria mentioned before, namely, of the fact that in conscious action we can 'take account' of all other processes which are also conscious.

6.12.   The same applies to another supposed criterium of consciousness which is often mentioned but to which we have not yet referred, namely, to the fact that conscious experience can be remembered and will be recognized as already experienced before when they occur again. 'Memory' or 'recognition' means here no more than the reappearance in consciousness, in combination with circumstances with which it has become associated, of what has been consciously experienced before. This is not a helpful criterium, if for no other reason so because it seems to be the case that we can sometimes in this manner 'remember' sense experiences of which we were not aware at the time when they first occurred. Moreover, memory in the general sense of learning clearly occurs also on pre-conscious levels.[1] With regard to conscious events the possibility of remembering or recognizing them thus means no more than that events which have occurred within the sphere of consciousness may do so again, but it tells us nothing about how conscious phenomena differ from non-conscious ones. All that it suggests is that in some sense all conscious phenomena belong to a common sphere, so that any conscious experience may appear in the company of any other .

6.13. The possibility of 'giving an account of' and of 'remembering' conscious events therefore merely leads us to that specially close connexion between all events which are at the same time conscious which before we have described by saying that in conscious processes a person will be able to 'take account' of other events which are conscious at the same time.

6.14. This close connexion between *all* conscious events, often described as the 'unity of the consciousness', can be regarded as a distinguishing attribute of conscious events, since the same does not seem to apply to unconscious mental events. While such unconscious mental processes which occur at the same time *may* also affect each other, this is not necessarily always the case. Even when they take place at the same time, they may proceed largely independently of each other (possibly in different subcentres) and without affecting each other's course. In other words, there is more than one 'unconsciousness' (or coherent system of unconscious mental events) while there exists (normally) in any individual only one consciousness.

6.15. Another familiar fact which is connected with this 'unity of the consciousness' may be mentioned here. This is the so-called 'narrowness of the consciousness', or the fact that, at any one time, only a limited range of experience can be fully conscious. Though the focus of consciousness may rapidly shift from one object to

[1]See above (5.10–5.12) and J. G. Miller, 1940.

another, this often seems to mean that processes which have been fully conscious may temporarily recede into a semiconscious or subconscious condition and persist there, ready to spring again into full consciousness at any moment.

3. THE COMMON SPACE-TIME FRAMEWORK

6.16. The 'unity of consciousness' means, above all, that conscious events occupy a definite position in the same spatial and temporal order, that they are 'dated' and 'placed' in relation to other conscious events, and that all sensory and affective events which 'enter consciousness', together with the reproductions or images of such experiences, belong to the same order or universe. This means that within the range of conscious events the 'generality' (as distinguished from the 'specificity', see 5.34 and 5.49) of their classification or evaluation has reached the highest degree: they are discriminated not only with respect to particular responses but with respect to all responses guided by conscious processes. This comprehensiveness of the system of relationships which connect and order all conscious events with respect to each other is probably the most characteristic attribute of these events.

6.17. This highest degree of comprehensiveness of the order of the system of relations prevailing on the conscious level does not necessarily mean that all the stimuli recorded by impulses proceeding at a lower level must also be capable of obtaining a distinct position at that higher level (4.53) or even that the number of different impulses which may be connected at that highest level must necessarily be larger than that of those which can possess connexions at some lower level. The comprehensiveness of which we speak is rather that of the top level of a hierarchic order at which all the elements belonging to that level are interconnected, while many elements belonging to lower levels may be connected with many other elements of the same level only *via* that higher level. It may be because all *classes* of impulses at this highest level form a common order, rather than because symbolic representations of all individual stimuli can reach this highest level, that it possesses this specially comprehensive character which distinguishes the system of relations on this level.

6.18. The existence of a common spatio-temporal framework, in which all the events which occur at that level are given a definite place, means that all reproductions or images of past or possible events will there be related to the experiences which are 'here' and 'now', and that this universal relatedness of all events to this common point of reference constitutes them into a continuum, the 'I'. The ever-presence of this common framework so long as consciousness is awake presupposes the continuous existence of certain representations of a most abstract kind: of a skeleton outline of the (spatial and temporal) surroundings within which we place the picture of the particular objects of which we are consciously aware or which we consciously imagine.

6.19. The continuous presence, while consciousness lasts, of these mental contents of a most abstract character, representing the spatial and temporal structure of the environment, is not always recognized. This is probably due to the preconception that concrete perceptions always precede the more abstract mental contents. There can be little doubt, however, that the distinct conscious picture of particular phenomena is always embedded in, or surrounded by, a semiconscious and more shadowy outline of the rest of the surroundings, which is co-present with, although much less distinct and detailed than, the conscious picture itself.

6.20. Of this co-presence of subconscious representations of the environment with the conscious representations of those parts of which we are clearly aware, we can easily convince ourselves if we remember the common experience of suddenly feeling to 'have the ground withdrawn under our feet' if something presupposed by the conscious picture proves to be missing. If, e.g., after our conscious sensations were interpreted on the assumption that we were in an enclosed room, we discover that the walls behind us are missing, or when, walking on what seemed to be level ground, we suddenly come to a precipice, this alters our view even of those things of which we were fully aware before. In all such instances the subconsciously presupposed background of our experience is discovered to be missing, and the firm 'placing' of the consciously experienced events in a subconsciously presupposed framework is upset. The result is usually a characteristic feeling of dizziness and of disturbed orientation.

6.21. Conscious experience thus rests on a much more extensive basis of less fully conscious or subconscious images of the rest of the surroundings, which nevertheless (like the following of a sensory impulse which determines its quality) give to the conscious representations their place and value. Conscious experiences have in this respect justly been compared to mountain tops rising above the clouds which, while alone visible, yet presuppose an invisible substructure determining their position relative to each other.

4. ATTENTION

6.22. Before we can attempt a more definite characterization of the peculiar attributes of conscious experience, it will be necessary to consider another phenomenon which is so closely related to consciousness that it must be regarded as little more than a specially high degree of awareness. This is 'attention'.[1] Our awareness of events to which we 'give our attention' certainly differs from that of other events of which we are merely co-conscious very much in the same manner in which the latter differ from conscious experiences.

6.23. The experiences to which our attention is directed are more fully discriminated and perceived in greater detail than others of which we are also aware. They stand out altogether more clearly than those which occupy merely the fringe of consciousness. We notice more in them and are more fully prepared to respond adequately to their occurrence.

6.24. At the same time, it is characteristic of attention that it has these effects only with regard to events which are in some sense expected or anticipated, and that, however attentive we may endeavour to be, an altogether unexpected event will take us as much by surprise and escape as much our detailed notice as if we had not been attentive at all. Attention is thus always directed,[2] or confined to a particular class of events for which we are on the look-out and which, in consequence, we perceive with greater distinctness when one of them occurs.

6.25. These characteristic attributes of attention fit readily into our account of the process of perception. Events or objects whose probable occurrence is suggested by the perceptions or images of which we are aware will form part of the following of these experiences. The corresponding physiological following will therefore

---

[1] Cf., E. G. Boring, 1933, pp. 231–2: 'Consciousness is attentive, attention is selective; consciousness is selective. Attention and consciousness are almost synonymous, and selection is the fundamental principle of both.'
[2] Cf., W. Stern, 1938, p. 474.

be in a state of excitatory preparedness, and this will facilitate the evaluation of any of the corresponding stimuli should they actually occur. The excitation produced by any such stimulus will as a result penetrate further, the network of relations determining its position will be more extensively activated, and the sensory impulse will therefore be more completely discriminated.

6.26. The fact that the following of certain impulses or groups of impulses, or certain parts of this following, will be more completely activated because the new impulses will be supported by an anticipatory stream of impulses tending in the same direction, will bring it about that these impulses will be evaluated more fully than others which may also be present; and the corresponding sensations will be lifted above others in intensity and distinctness.

6.27. This kind of anticipatory excitation of parts of the following of certain kinds of sensory impulses will mean that we shall not only be more ready to perceive the corresponding stimulus, but also that we shall perceive them from a certain angle or a certain 'point of view'; we shall discriminate them more fully with respect to certain types of responses towards which the whole organism is disposed at the moment. The conception of 'disposition' or 'set' which we have discussed before in connexion with 'purposiveness' (4.60–4.63), is indeed the most general manifestation of the process of which attention is a special instance.[1]

6.28. It is worth noting in this connexion that something similar to attention can be observed also on a semiconscious or subconscious level. A person may be interested in finding some object or in noticing a particular event, and, although not thinking about it, will at once observe it when it presents itself to his senses, merely because his mind has been prepared for it. The same is probably true of certain suggestions in a hypnotic state which are carried out when the occasion arises. These states of preparedness for certain actions, like attention proper, or the set or disposition corresponding to a particular urge or drive, must probably be conceived as determined by a state of excitatory preparedness of the following of the classes of impulses corresponding to the objects to which they are directed.

[1]Cf., J. G. Miller, 1942, p. 159: 'Closely related to attention, perhaps its outward manifestation, is the phenomenon of set'.

5. THE FUNCTIONS OF CONSCIOUSNESS

6.29. The phenomenon of attention is of special significance for the understanding of consciousness because in the case of attention as well as in that of consciousness, it is not merely the character of the particular stimulus, or the place of the impulse in the network of connexions, which will determine whether it will or will not become conscious or receive our attention; but it will be the pre-existing excitatory state of the higher centres which will decide whether the evaluation of the new impulses will be of the kind characteristic of attention or consciousness. It will depend on the predisposition (or set) how fully the newly arriving impulses will be evaluated or whether they will be consciously perceived, and what the responses to them will be.

6.30. It is probable that the processes in the highest centres which become conscious require the continuous support from nervous impulses originating at some source within the nervous system itself, such as the 'wakefulness centre' for whose existence a considerable amount of physiological evidence has been found.[1] If this is so, it would seem probable also that it is these reinforcing impulses which, guided by the expectations evoked by pre-existing conditions, prepare the ground and decide on which of the new impulses the searchlight beam of full consciousness and attention will be focused. The stream of impulses which is thus strengthened becomes capable of dominating the processes in the highest centre, and of overruling and shutting out from full consciousness all the sensory signals which do not belong to the object on which attention is fixed, and which are not themselves strong enough (or perhaps not sufficiently in conflict with the underlying outline picture of the environment) to attract attention.

6.31. There would thus appear to exist within the central nervous system a highest and most comprehensive centre at which at any one time only a limited group of coherent processes can be fully evaluated; where all these processes are related to the same spatial and temporal framework; where the 'abstract' or generic relations form a closely knit order in which individual objects are placed; and where, in addition, a close connexion with the instruments of communication has not only contributed a further and very powerful means of classification, but has also made it

[1] Cf., C. T. Morgan, 1942, p. 283 *et seq.*

possible for the individual to participate in a social or conventional representation of the world which he shares with his fellows.

6.32. Of these various characteristics of consciousness the predominance of 'abstract' features requires further discussion. This will be combined in the next two sections with some further examination of the nature of abstract thought.

6. 'CONCRETE' AND 'ABSTRACT

6.33. While the 'consciousness' of consciously experienced qualities implies their being closely related to each other, the relationships which determine these qualities are not in turn themselves conscious. These relationships determine how different conscious experiences will act upon or affect each other, but they are present in consciousness only in this 'implicit' manner and are not explicitly experienced.

6.34. Since the relations which determine the character of mental qualities are not themselves consciously experienced but show themselves solely in the different effects which the different experienced qualities produce, the latter appear to us as the absolute and irreducible data of consciousness. This is often expressed by the statement that sensory experience gives us acquaintance with the 'concrete' phenomena while the higher mental processes derive 'abstractions' from those immediate data.

6.35. This distinction between the 'concrete' character of immediate experience and the 'abstract' character of 'concepts' is misleading in several respects. It is closely connected with the old belief that the sensory qualities constitute in some sense a reproduction of corresponding attributes of the objects of the external world, and with that mosaic theory of perception which conceives of all mental events as being built up from fixed sensory 'elements'.

6.36. If sensory perception must be regarded as an act of classification, what we perceive can never be unique properties of individual objects but always only properties which the objects have in common with other objects. Perception is thus always an interpretation, the placing of something into one or several classes of objects. An event of an entirely new kind which has never occurred before, and which sets up impulses which arrive in the brain for the first time, could not be perceived at all.[1]

[1]Cf., H. Henning, 1924, p. 304.

6.37. All we can perceive of external events are therefore only such properties of these events as they possess as members of classes which have been formed by past 'linkages'. The qualities which we attribute to the experienced objects are strictly speaking not properties of that object at all, but a set of relations by which our nervous system classifies them[1] or, to put it differently, *all* we know about the world is of the nature of theories and all 'experience' can do is to change these theories.[2]

6.38. This means also that what we perceive of the external world are never either all the properties which particular objects can be said to possess objectively, nor even only some of the properties which these objects in fact do possess physically, but always only certain 'aspects', relations to other kinds of objects which we assign to all elements of the classes in which we place the perceived objects. This may often comprise relations which objectively do not at all belong to the particular object, but which we merely ascribe to it as the member of the class in which we place it as a result of some accidental collocation of circumstances in the past.

6.39. In any sense in which we can contrast our knowledge of the perceived properties of an external object with its physical or objective properties, all sensory perception is therefore in a sense 'abstract'; it always selects certain features or aspects of a given situation. We shall presently see that the assumption that there exists a physical world different from the phenomenal world involves the assumption that the former possesses properties which we cannot directly perceive, and even some which we do not know. Even the so-called elementary sensory qualities are in this sense 'abstractions', since they are determined by bundles of relationships which we have learnt to attach to certain stimuli which in a physical sense may or may not possess identical properties.

6.40. We already have stressed repeatedly the fact that the immediate data of consciousness are not in fact built up in mosaic fashion from elementary sensations. We perceive directly such complexes as configurations (gestalts), and there can be little

---

[1]Cf., E. G. Boring, 1933, p. 30: 'The thesis of this book is that nothing is "directly observed", that every fact is an implication.' It is curious that an author holding this view should still describe himself as a positivist.

[2]I owe this way of putting it to my friend K. R. Popper, who, however, may not entirely agree with this use I am making of his ideas.

doubt that we often consciously perceive only the gestalt qualities without being aware of the 'elementary' sensations (such as colours) of which the former were once supposed to be built up.[1] It is at least probable that even on a pre-conscious level we can learn to respond similarly to certain 'abstract' features of an external situation irrespective of the different elements of which the gestalt may be built up in the particular case.

6.41. The immediate data of consciousness will therefore be 'abstract' not only in the sense that they can never convey to us more than certain generic 'attributes' of the perceived objects, but also in the sense that they will always reflect only some of these generic properties which might be ascribed to the perceived object. If, with regard to current perceptions, we are usually little aware of their partial or incomplete character, this is probably due to the fact that while the experience is present we are in a position to supplement it by directing our attention to particular features.

6.42. This possibility of 'filling in' at first unperceived details by directing our attention to them probably constitutes also one of the main differences between current perceptions and memory images. (Though some people of the eidetic type appear to be able by recalling vivid images to discover details in them which they had not noticed at the time of the original experience). But the memory images need not always to be more 'abstract' than current perceptions. If frequently only certain abstract features of a perceived situation can be remembered, this is maybe a consequence of the fact that only those abstract features were perceived in the first instance.

6.43. While there thus exists little justification for any sharp distinction between the 'concrete' picture supplied by sense perception and the 'abstractions' which are derived from the former by the higher mental processes (or between the complete picture of a unique situation built up by the 'senses' from fixed elements, and the abstract features which the 'intellect' singles out from the picture which is supposed to be given prior to any abstraction, cf. 5.19), there is a legitimate sense in which we can at any moment distinguish between the immediate data of consciousness and the further processes of rearrangement and reclassification to which they can be subjected on a conscious level.

---

[1] J. von Kries, 1923, p. 99.

7. CONCEPTUAL THOUGHT

6.44 We have seen that the classification of the stimuli performed by our senses will be based on a system of acquired connexions which reproduce, in a partial and imperfect manner, relations existing between the corresponding physical stimuli. The 'model' of the physical world which is thus formed will give only a very distorted reproduction of the relationships existing in that world (5.20–5.24); and the classification of these events by our senses will often prove to be false, that is, give rise to expectations which will not be borne out by events.

6.45. But, although the conscious mind can know of the external world only in terms of the classes which earlier experience has created, and although all its conscious experience must always refer to elements of such given classes rather than to individual objects, the experience of these data of consciousness will provide the foundation for a revision of the classification from which it starts. Further experience will show that parts of different situations which our senses represent as being alike will, according to the different accompanying circumstances, have to be treated as different. The mind will perform on the initial sensory experiences a process of reclassification, the object of which are no longer the original stimuli but the elements of the classes formed by the pre-conscious sensory mechanism.

6.46. The experience that objects which individually appear as alike to our sense will not always behave in the same manner in relation to other classes of apparently similar objects, and that objects which to our senses appear to be different may in all other respects prove to behave in the same manner, will thus lead to the formation of new classes which will be determined by explicitly (consciously) known relations between their respective elements. These new classes formed by a rearrangement of the objects of the sensory world are what are usually described as abstract concepts.

6.47. The formation of abstract concepts thus constitutes a repetition on a higher level of the same kind of process of classification by which the differences between the sensory qualities are determined.[1] This continuous process of reclassification is forced on us because we find that the classification of objects and events which our senses effect is only a rough and imperfect approximation

[1] Cf., H. Margenau, 1950, pp. 54–56, and H. Werner, 1948, p. 222, 224.

to a reproduction of the differences between the physical objects which would enable us correctly to predict their behaviour[1] —an approximation determined by the accidents of evolution, the physiological capacities and the pragmatic needs of the individual and the species.

6.48.   Perhaps we may go still further and regard conceptual thought and the processes of inference as a further repetition of the process of classification on a still higher level. It is probably no accident that the formation of classes and the relation between classes were first studied in the attempt to analyse the principles of conceptual reasoning. It should be clear now that the same kind of relationship which in logic has been developed as the theory of classes and relations, is immediately applicable to that physiological process of multiple grouping or classification which we have been examining. And it should not be difficult to conceive of the conscious mental process which logic analyses as a repetition on a higher level of similar processes which, on a pre-conscious level, have produced the material on which the conscious processes operate.

6.49.   We cannot attempt here further to distinguish the different levels on which this kind of process of constantly repeated classifications proceeds, and we must be content with the suggestion that all the 'higher' mental processes may be interpreted as being determined by the operation of the same general principle which we have employed to explain the formation of the system of basic sensory qualities. We have throughout to deal with an ever-repeated process of classification of the kind described earlier.

6.50.   With this suggestion of the essential unity of the character of the physiological mechanism underlying all kinds of mental processes we have concluded the exposition of the theory which is the main object of this study. There remain yet, however, two complementary tasks to which the two concluding chapters will be devoted. We shall first consider what kind of empirical confirmation or refutation we may hope to find for our theory. The final chapter will examine certain philosophical consequences which would follow from this theory and which are closely connected with certain topics merely touched upon in this and the preceding chapters.

----

[1]Cf., F. A. Hayek, 1942, p. 271 *et seq.*

CHAPTER VII

CONFIRMATIONS AND VERIFICATIONS OF THE THEORY

1. OBSERVED FACTS FOR WHICH THE THEORY ACCOUNTS

7.1.   The value of a theory of the kind presented here may prove itself by accounting for known facts as consequences of other known phenomena, by enabling us to eliminate phantom problems, by showing that certain earlier theories are special cases of a more general principle, or, finally, by suggesting new questions which can be experimentally investigated. On all these scores our theory appears to have a certain amount of *prima facie* evidence in its favour.

7.2.   The main aim of the theory presented is to show that the range of mental phenomena such as discrimination, equivalence of response to different stimuli, generalization, transfer, abstraction, and conceptual thought, may all be interpreted as different forms of the same process which we have called classification, and that such classifications can be effected by a network of connexions transmitting nervous impulses. From the fact that this classification is determined by the position of the individual impulse or group of impulses in a complex structure of connexions, extending through a hierarchy of levels, follow certain important conclusions concerning the effects which physiological or anatomical changes must be expected to have on mental functions. We shall confine ourselves here to point out a few of the more important consequences of our theory which are in accordance with observed fact.

7.3.   Since the qualities of mental events produced by particular impulses or groups of impulses according to this view depend not on any property which these impulses possess by themselves, but on their position in the whole network of connexions, it would follow that the different mental functions need not be localized in any particular part of the cortex.

7.4. While the possibility of a peripheral stimulus producing a sensory quality will in general[1] depend on the preservation of the central endings of the corresponding afferent fibres, there is no reason for expecting that beyond this the capacity of experiencing particular qualities will depend on particular parts of the cortex. We should rather expect to find, as in fact we do find, that a destruction of a limited part of the cortex will lead to some weakening of most or all mental functions, rather than to the extinction of some particular capacities.[2]

7.5. Similar considerations would lead us also to expect that particular mental functions will not depend entirely on the existence of particular nervous connexions but will be capable of being produced by alternative channels. If the complete classification which determines the peculiar mental quality of an impulse depends on a multiplicity of connexions extending throughout the greater part of the cortex, this does not mean that for any particular effect any one of these connexions will be indispensable. Partial classifications based on certain bundles of connexions may often alternatively be capable of bringing about a discrimination sufficient to maintain the particular effect.

7.6. This may mean either that certain mental processes which are normally based on impulses proceeding in certain fibres may, after these fibres have been destroyed, be relearned by the use of some other fibres, or that certain associations may be effectively brought about through several alternative bundles of connexions, so that, if any one of these paths is severed, the remaining ones will still be able to bring about the result. Such effects have been observed and described under the names of 'vicarious functioning' and 'equipotentiality'.[3]

7.7. Our account of the translation of the neural impulse into a mental event as a process of classification leads us to expect that we will find that this process not only takes perceptible time but also that it can be observed in different successive stages in which the classification or evaluation is developed to different degrees.

[1] In view of what has been said before (4.37–4.42) about the rôle of the low-level responses it is, however, not entirely inconceivable that in the case of a local destruction of the cortical endings of particular sensory fibres, the proprioceptive fibres recording short-arc responses may come to deputize for them.

[2] See K. S. Lashley, 1929, and already J. von Kries, 1898.

[3] K. S. Lashley, 1929.

This expectation is amply borne out by observation. From the unconscious responses to stimuli and the still unconscious 'subception'[1] through the 'pre-sensation'[2] and the various degrees of clarity of the sensation,[3] perception and 'apperception', to judgements and concept formation, there exists clearly a chain of events in which the full evaluation of any mental quality gradually unfolds itself.

7.8. From the account we have given of the determination of sensory qualities it would further follow that the quality of any sense experience attached to certain impulses or group of impulses will not always be the same but will be different in different circumstances. The same individual stimulus, affecting the same receptor organs, must thus be expected to produce different sensory qualities according as different other stimuli operate at the same time.

7.9. As we have already seen (4.45–4.47), this expectation is fully borne out by experimental work. Many stimuli are perceived 'correctly' only if received under normal conditions, but lead to different sensations if the setting is not normal.[4]

2. OLDER THEORIES COMPRISED AS SPECIAL CASES

7.10. There is no need here to mention again the various instances where our approach eliminates what now appear to be false questions. We can at once turn to the several instances where our theory embraces as special cases theories which in the past have been advanced in order to explain particular phenomena. Some of these instances have been noticed earlier and now need be mentioned only briefly.

7.11. The first instance of this kind which has been discussed earlier (3.40–3.45) is Berkeley's theory of spatial vision and the more general theories of space perception which have developed from it. The account of the determination of the spatial order of perception by the co-ordination between the various sense modalities and the kinesthetic sensations is of course merely one particular

---

[1]R. A. McCleary and R. S. Lazarus, 1949, p. 178.
[2]F. R. Bichowski, 1925, p. 589, R. B. Cattell, 1930.
[3]H. Henning, 1922, p. 71.
[4]See on this now the German works by W. Metzger, 1940, and V. von Weizsaecker, 1941.

instance of the theory of the determination of sensory qualities developed here.

7.12. Another similar instance of an anticipation in a particular field which we have already mentioned is the James-Lange theory, of emotions. As has been shown before (4.70–4.72), this theory, carefully restated, might be regarded as a special case of our theory.

7.13. In the case at least of von Helmholtz the emphasis on the effect of experience in determining sensory qualities goes far beyond ascribing to experience the creation of their spatial order, and it probably is due mainly to his influence that it is to-day widely recognized that 'the manner in which we see things of the external world is sometimes affected by experience to an overwhelming extent' and that 'it is often difficult to decide which of our visual experiences are determined immediately by sensation and which, on the contrary, are determined by experience and practice.'[1] His conception of the 'unconscious inference',[2] by which stimuli which do not lead to conscious experience are yet utilized in the perception of a complex position, comes very close to the theory developed here. Yet von Helmholtz, like all later writers following on these lines, instead of drawing the conclusion that the factors to which he attributed 'overwhelming importance' in determining the sensory qualities might be the sole factors which determine them, in fact insisted that nothing could be recognized as sensation which is demonstrably due to experience[3]—thus giving, in fact, support to the conception of a pure core of sensation.

7.14. The same applies to the group of theories which have furthest developed this line of thought, the *Reproduktionspsychologie* of B. Erdmann, R. Dodge, H. Henning and F. Schumann, which, with its stress on the 'residua' which determine sensory qualities, came very close to the position taken here, yet never ceased to distinguish between a 'stimulus component' and a 'residual

[1] I cannot now trace the source of this quotation, but similar statements can be found in many passages of Helmholtz, e.g., 1866 (1925) III. p. 12. W. Wundt's theory of 'assimilation,' which ought also to be mentioned in this connexion, is essentially a development of these ideas.

[2] H. von Helmholtz 1866 (1925) III, p. 4, where *'unbewusster Schluss'* is, however, inadequately translated as 'unconscious conclusions'. The correct translation 'unconscious inference' is suggested by E. G. Boring.

[3] H. von Helmholtz 1866 (1925) III, p. 13.

component', the former of which still corresponds to the 'pure core' of sensation.[1]

7.15. The relation which exists between our theory and the views of the gestalt school is of a somewhat different character and has already been discussed (3.70-3.79). As was then pointed out, the present approach may be regarded as an attempt to raise, with regard to all kinds of sensory experiences, the question which the gestalt school raised in connexion with the perception of configurations. And it seems to us, that in some respects at least, our theory may be regarded as a consistent development of the approach of the gestalt school.[2]

7.16. Another instance of a connexion between our theory and a familiar older view has not yet been explicitly mentioned: the obvious relations which exist between it and the basic ideas of the old association psychology. Our view agrees, of course, with associationism in the endeavour to trace all mental processes to connexions established by experience between certain elements. It differs from it by regarding the elements between which such connexions are established as not themselves mental in character but as material events which only through those connexions are arranged in a new order in which they obtain the specific significance characteristic of mental events (5.52).

7.17. This is a step which James Mill very nearly made when he briefly suggested that similarity ('resemblance') might be dispensed with as a 'principle of association' and be reduced to a 'particular case' of the 'law of frequency' of co-occurrence.[3] This promising beginning was, however, cut short by the somewhat uncomprehending comment added to this passage by his son, who described the brief hint as 'perhaps the least successful attempt at simplification and generalization of the laws of mental phenomena, to be found in the work.' The only further development of this idea is to be found in the writings of the last of the old

[1]B. Erdmann, 1886, 1907 and especially 1920, pp. 7, 16, 18, 63-64, 74-75, and 127; R. Dodge, 1931, p. 126; H. Henning, 1917, p. 198, 1924, pp. 303-304; F. Schumann, 1908, II, p. 19, 1922, pp. 207, 216. It may be worth mentioning that the fullest exposition of this view, Erdmann's *Reproduktionspsychologie*, appeared in the same year, 1920, when the first draft of the present theory was completed.

[2]This applies particularly to the formulation of the basic problems by K. Koffka, 1935.

[3]J. Mill, 1829 (1869), I, p. 111.

association psychologists, G. H. Lewes, which never seem to have received the attention which they deserve.[1]

7.18. Finally, we may perhaps once more mention that within the framework of this theory the conception of events which are mental but not conscious receives, for the first time so far as we are aware, a clear meaning. In consequence it provides a systematic place for whatever of the various theories of the unconscious will prove permanent additions to knowledge.

3. NEW EXPERIMENTS SUGGESTED

7.19. The theory developed here is not the kind which one could hope to confirm or refute by a single crucial experiment. Its value ought to show itself rather in suggesting new directions in which experimental work should produce interesting results. The main thesis for which one may hope to find experimental confirmation is that the sensory qualities can be changed by the acquisition of new connexions between sensory impulses. If this central contention is correct it should in principle be possible both, to attach conscious sensory qualities to sensory impulses which before carried no conscious values, and to create discriminations between such impulses which before caused undistinguishable sensations. It should even be possible to create altogether new sensory qualities which have never been experienced before.

7.20. There exists a great deal of evidence that the capacities for sensory discrimination can be greatly developed by practice. The greatly heightened capacities for tactual, auditory and olfactory discrimination often acquired by the blind,[2] the development of taste, smell, vision and touch by the professional tasters and samplers of wine,[3] spirits, tobacco, chocolate, perfumes, wool,[4] cheese,[5] and the like, the development of the sense of smell by

---

[1]G. H. Lewes, 1879.

[2]In addition to such older studies as the classic investigation of J. N. Czermak, 1855, and the more recent works of J. T. Williams, 1922, and M. von Senden, 1932, see the recent summary by E. von Skramlik, 1937, p. 173, which seems to show that the predominant evidence is against the contrary results obtained by some investigators.

[3]H. Henning, 1924, p. 55.

[4]H. Binns, 1926.

[5]G. W. S. Blair and F. M. V. Coppen, 1939.

some doctors and chemists,[1] of the auditory sense of musicians[2], and of the colour sense of artists and dyers[3] are familiar, although quite inadequately studied, examples.

7.21. In more recent times, largely under the influence of the gestalt school, the effect of experience and practice on what has come to be known as 'perceptual organization' has received a good deal of attention. It appears to have been established beyond doubt that the perception of the various configurations and complexes can be profoundly altered by experience.[4] But although this fact is closely connected with our problem, and (if the belief, held both by the gestalt school and ourselves, is correct, that there is no real difference between sensation and perception) goes far to make the variability of even the most elementary sensory qualities probable, it does not directly confirm that the latter is the case.

7.22. Most of this discussion of sensory organization—not excluding much of the work of the gestalt school, in spite of its fight against the 'constancy assumption'—however, still suffers from an underlying belief that this problem is one of how given sensations become 'organized', as if there could be unorganized sensory data, something like W. James's 'blooming buzzing confusion' in the mind of the newly born, and that it is these initial fixed sense data which perception organizes in a pattern.[5] These remnants of the old 'mosaic theory' which still pervade the discussion cannot be finally eliminated until it is realized that sensory organization and the determination of the individual qualities are one and the same problem.

7.23. Connected with the studies of the effect of experience on sensory organization are the known facts about the manner in which congenitally blind who by an operation have become able to see (and animals reared in darkness)[6], learn to perceive visual

[1]R. W. Moncrieff, 1942, pp. 9, 76.
[2]F. L. Dimmick, 1946, p. 19.
[3]E. G. Boring, 1942, pp. 339–340.
[4]K. W. Braly, 1933; R. W. Leeper, 1935; K. Duncker, 1939.
[5]As R. S. Woodworth rightly points out with regard to form perception (1938, p. 624), 'the empiricist theory aims to get along with a minimum number of concepts: it uses only the concept of a pure mosaic of elementary sensation and the concept of associations established by experience. To the associations are assigned the functions (a) of combining the elements into forms and (b) of giving objective meaning to these forms.'
[6]A. H. Riesen, 1947.

objects. The ample material collected on this problem by Senden[1] shows clearly that at least the ordering of the individual sensations has to be gradually learnt, but also that apparently such persons are able from the first moment to distinguish colours. But as it appears that no completely blind person has ever gained vision in this manner[2] and that all those operated persons whose vision had been obstructed by cataract were, before the operation, able to distinguish shades of light and probably also colours, this information is of little direct use for our purpose.

7.24. Perhaps the most significant experimental findings in this field are the extensive investigations of Stratton, Ewert and, more recently, Erismann on the effect of the prolonged wearing of various kinds of spectacles which either invert or distort vision,[3] and the corresponding experiments by P. T. Young with the 'pseudophone', an apparatus which effects an acoustical transposition of sound between the two ears.[4] All these experiments show that the significance or position of different stimuli of one modality relative to stimuli of another modality can be altered if they are regularly made to occur in a new combination.

7.25. The older treatises on psychology contain a good deal of discussion on the effect of practice on sensory discrimination. William James, e.g., in a section headed 'the improvement of discrimination by practice' even explicitly mentions as the first cause 'which we can see at work whenever experience improves discrimination' the fact that 'the terms whose difference comes to be felt contract disparate associates and these help to draw them apart'.[5]

7.26. Little systematic work, however, has been done on this problem and even the meaning of the concept of practice as applied to sensory discrimination, and of the conception of new or

---

[1]M. von Senden, 1932.
[2]J. B. Miner, 1905, p. 103.
[3]G. M. Stratton, 1897; P. H. Ewert, 1930 and 1936; T. Erismann, 1948.
[4]P. T. Young, 1928.
[5]W. James, 1890, I, pp. 508, 510. For other similar earlier references *see* O. Külpe 1895, pp. 42, 302, 340; L. J. Martin and G. E. Müller, 1899, pp. 128ff; E. B. Titchener, 1905 I,ii, p. 57; E. L. Thorndike, 1913, p. 152; F. Krueger, 1915, pp. 95–96; J. von Kries, 1923, p. 144. The explicit and categorical denial of any improvement of the capacity of sensing, or sensory acuity, by H. L. Kingsley, 1946, p. 265, is rather exceptional and apparently based on as little precise information as the prevalent contrary view.

improved discriminations, has been left somewhat obscure. Indeed the older psychologists who paid at least some attention to the effect of practice in this connexion, were inclined to regard it mainly as a nuisance, an effect which had to be eliminated before serious experimental work could start, rather than as a phenomenon which deserved investigation for its own sake.

7.27.   The earliest and for a long time the only systematic experiments in these fields were those performed nearly a hundred years ago by A. W. Volkmann[1] on the effect of practice on the threshold for discrimination between two neighbouring points on the skin. Later experiments[2] have amply confirmed his findings that not only these thresholds could be decreased by short practice by as much as from 50 mm. to 0.5 mm., but also that practice with such tactual stimuli on a part of the skin on one side of the body would similarly decrease the threshold for discrimination between symmetrically corresponding points on the other side of the body.

7.28.   Almost the only systematic work done in this field in more recent times are a number of somewhat inconclusive studies on the effect of practice on pitch discrimination in hearing, conducted by various students interested mainly in musical education.[3] These studies are not very helpful for our purpose because they addressed themselves in the main to the question whether practice would improve discrimination, rather than to the problem of the conditions under which it would do so. The one significant point which emerges is that it seems to be generally true that no mere repetition but only knowledge of results of the attempts to discriminate will lead to an improvement of discrimination.

7.29.   This unsatisfactory state of knowledge of the whole subject is probably in a great measure due to the uncertain meaning of the concept of practice when applied to these problems. Although this meaning is usually taken (and sometimes explicitly said[4]) to be obvious, it is by no means clear that the sense of improving an existing 'capacity' by repeated exercise, which is probably

[1]A. W. Volkmann, 1858.
[2]See V. Henri, 1898, and the summary of the work of F. B. Dressler, G. A. Tawney, and L. Solomons in C. L. Friedline, 1918.
[3]H. T. Moore, 1914; J. F. Humes, 1930; A. A. Capurso, 1934; E. Connette, 1941; C. H. Wedell, 1945; R. Wyatt, 1945; B. L. Ricker, 1946.
[4]E.g., B. J. Underwood, 1949, p. 118.

roughly what is meant by the effect of practice in other fields, fits the case of sensory discrimination.

7.30. There is little difficulty about understanding why the repetition of any particular series of movements should enable us to perform them afterwards more quickly, surely, smoothly or otherwise more efficiently. But there seems to be no similar obvious reason why any number of attempts to distinguish between two stimuli which we have not been able to distinguish before should teach us to do so. The whole approach to the problem seems still to be determined by the rather meaningless conception that these different sensations are always 'there' in some concealed sense, and that the problem is merely to learn to notice these 'unnoticed' sensations which are assumed to be necessarily and invariably coupled with the sensory impulse.

7.31. With regard to any kind of movements, practice clearly means some effect of memory and, as we have seen (5.10–5.12), it is difficult to see what other meaning 'memory' can have but the retention of connexions or relations. But while this conception applies directly to the acquisition of new series of movements which can become coupled with each other, and makes it easy to see why, e.g., such a series of movements which at first could be performed only by conscious effort, comes later to be performed automatically, at least the traditional view of the character of sensations does not fit into this pattern.

7.32. To acquire the capacity for new sensory discrimination is not merely to learn to do better what we have done before; it means doing something altogether new. It means not merely to discriminate better or more efficiently between two stimuli or groups of stimuli: it means discriminating between stimuli which before were not discriminated at all. If qualities are, as we have maintained, subjective, then, if new discriminations appear for the first time, this means the appearance of a new quality. There is no sense in saying that, if a chemist learns to distinguish between two smells which nobody has ever distinguished before, he has learnt to distinguish between given qualities: these qualities just did not exist before he learnt to distinguish between them.

7.33. Of course, such a 'new' quality can never be unlike any quality ever experienced before: to be recognizable as a distinct quality it must, in certain ways, be related to already familiar qualities, be in various respects similar to, or different from them.

156

It will be a quality only by occupying a certain position in the order of all qualities, an order which can only be gradually extended and more finely subdivided. But although thus most 'new' qualities will constitute merely a new step in a pre-existing gradation or scale, and share their various attributes with different other qualities, they will nevertheless be new qualities which did not exist before.

7.34. The prevalent uncritical attitude towards the whole problem probably has been much assisted by the fact that the very term 'discrimination' suggests something like a 'recognition' of objective differences between the stimuli (2.32) and belongs thus to an earlier stage of theoretical development.[1] To this idea probably is also due the still widely held view that what is affected by practice is merely the 'interpretation' of a 'given' sensory quality or datum. The whole problem is still largely approached as if the differences between sensory qualities could be accounted for by a different physiological sensitivity of the sense organs—a physiological 'capacity' which needs merely to be 'developed' and which at the same time sets a 'physiological limit' to the extent to which discrimination can be improved. These concepts of the 'capacity' and the 'physiological limit' are as obscure and need as much clearing up as the concept of practice itself.

7.35. Discrimination in the relevant sense (better described as classification) involves not only the learning to respond differently to different physical stimuli, but also the learning to respond similarly to stimuli which physically may be different or similar, and to respond differently to the same stimulus in different contexts. In order that a problem of discrimination should arise, it is necessary, of course, that the different stimuli should cause impulses in different sensory fibres (or, though this does not seem to be the case, different kinds of impulses in the same fibre). But this condition would appear to be the only 'physiological limit': different impulses which affect the same receptor organs in the same manner must under the same conditions produce undistinguishable effects.

[1] It is interesting to note that E. G. Boring, who at one stage had defined consciousness as discrimination (1933, p. 187), later came to the conclusion that it is 'probably best to abandon the word *discrimination*, which implies a freely acting, conscious observer, and to limit ourselves to the descriptive terms of successive differentiation and relations between them'. (1937, p. 451.)

7.36.   Unless we assume the theory of the specific energy of the nerves to be true in its illegitimate interpretation (1.33), there is indeed no reason why it should not be possible to learn to attach different qualities to impulses caused by stimuli which are physically identical, and proceeding in fibres which belong to the same sense modality. Cases are, of course, known where identical physical stimuli, acting on receptors belonging to different modalities ('paradoxical cold', vibration and sound, and the different sensory qualities produced by the same physical stimulus acting on the mucous membranes of the eye and the mouth—1.40) produce different sensations, but the same should in principle also be possible where otherwise identical receptors at different points of the body are involved.

7.37.   From the whole approach followed in the present inquiry it would follow that learning to distinguish between different individual stimuli can only mean that we come to attach to these stimuli different effects irrespective of the manner in which these stimuli differ objectively. Learning to discriminate does not necessarily produce a better reproduction of the physical order of the stimuli; it merely means the creation of a new distinction in the phenomenal order which, if it were the result of a non-recurring, accidental or artificial combination of stimuli during a particular period, might indeed prove later not a help but an obstacle to orientation and appropriate behaviour.

7.38.   The only sense in which the improvement of sense discrimination by practice can be said to be a 'development' of pre-existing capacities is that, in order that such discrimination at the higher levels should become possible, the occurrence of distinct processes at some lower level (at least the receptor level) must be presupposed. That is, the organism must initially respond in some way differently to the different stimuli (even if it only be that impulses are set up in the first instance in different fibres) if it is to be possible that these stimuli should acquire different significance for the higher nervous centres. It is at least likely that in most instances different responses to the impulses in the different fibres will already have taken place on a reflex or spinal level before the higher centres learn to discriminate between those impulses, since the development of distinct receptors for different physical stimuli probably goes hand in hand with the development of different responses to those stimuli.

7.39. There appear to exist three principal ways in which the attaching of new connexions to sensory impulses which arrive at the higher centres might lead to the appearance of new sensory qualities: 1. impulses which before did not produce a distinct sensation might come to do so; 2. different impulses produced by different physical stimuli which formerly produced the same sensory quality might be made to be perceived as different sensory qualities; and 3. impulses produced by the action of physically identical stimuli on similar receptor organs at different points of the body might also acquire different sensory qualities.

7.40. The task of experimentation in all these instances would be to ascertain whether we could either become aware of particular sensory impulses of which we were before not conscious, or whether sensory impulses could be given distinct qualitative significance different from that of other such impulses from which they were formerly indistinguishable; this might be done by attaching to them a distinct set of connexions which are different from those attached to other such impulses which before were perceived as identical.

7.41. It would seem that in any such experiments we must be able to rely on verbal reports of the subject and that therefore animal experiments cannot be used for our purpose. It would be necessary to ascertain before experiments start that the subject is either unaware of the stimulus, or unaware of any qualitative difference between the effects of different stimuli. And although we can teach animals to discriminate between stimuli with respect to certain responses, it would be impossible to decide whether an animal has merely learnt to attach a new response to distinct sensations which it perceived before, or whether it has acquired a new capacity for discrimination. Considering the difficulty of merely ascertaining, e.g., whether particular animals possess colour vision or not,[1] it would seem that animal experiments must be ruled out in this connexion.

7.42. With human subjects the chances of successful experiments on these lines probably differ greatly between the different sense modalities. With a sense as highly developed and as fully used in humans as sight, practice in most instances probably has been carried to a point where a definite order has been so deeply engrained that it would at least take very long to obtain any results.

[1]G. L. Walls, 1942, p. 472.

It should be noted, however, that as von Kries has pointed out,[1] in another sense this most highly developed of human sensory capacities is the most imperfect of the senses: the correspondence between physical differences between the stimuli and the differences between the sensory qualities is probably less close here than it is with other senses. Every colour can be produced by a great variety of mixtures of wave lengths in addition to (in most instances) a monochromatic (homogeneous) light. We do not know whether this equivalence of various combinations of stimuli is determined by a peripheral (i.e. receptor) or by a central mechanism. If the latter were the case, it should not be impossible to learn to see as different colours different mixtures of light waves which initially appear to be indistinguishable.

7.43. Better chances of experimental results exist probably in the less practised sense modalities, particularly those, such as the human sense of smell, of which in an earlier state of development man made greater use than he does in civilized life, and where the physiological capacities of distinguishing between different stimuli is probably much greater than that which we use. It has been pointed out by a competent observer that in this field 'the influence of practice is so enormous, particularly in the beginning, that some people require on the second day of experimentation only small fractions of the threshold values necessary on the first day, and that they then easily solve qualitative analyses which on the previous day seemed impossible to them.'[2]

7.44. As against this advantage of the relative unpractised state of olfaction in civilized men, and the consequent high degree of educability of this sense, stands our ignorance of the nature of the proximal stimuli[3] and of the differential sensitivity of the receptor organs for these stimuli. We shall nevertheless outline the kind of experiments which might be attempted with respect to this sense as the one which seems on the whole to be the most promising one for our purposes.

7.45. The task of the experiment would be to attempt to attach to originally undiscriminated stimuli as many distinct connexions with other sensory stimuli and emotional states as possible. That such intersensory associations can be created has, of course, been

[1] J. von Kries, 1923, p. 80.
[2] H. Henning, 1927, p. 745.
[3] See, however, the reports on recent work by L. H. Beck and W. H. Miles, 1947.

demonstrated by the recent work on sensory conditioning.[1] The problem is whether by attaching such distinct associations to initially indistinguishable stimuli new discriminations can be created.

7.46. Experiments had probably best start with stimuli which highly practised persons are known to distinguish, but which to an unpractised person are undistinguishable. The points to be ascertained would be not only whether by repeated exposure to the stimuli people can be taught to discriminate between them, but whether this process is considerably speeded up if the different stimuli are made to act under completely different accompanying circumstances. This implies of course the necessity of parallel control experiments in which the conditions under which the two stimuli act are the same.

7.47. For such experiments it would be desirable to alter the whole surroundings and the state of the organism in which the different stimuli were made to act: one of two stimuli might, e.g., be made to act regularly at a particular time of the day (say on awakening in the morning) so that it always coincided with the same phase of the rhythm of the body, in a state of restedness, warmth and inactivity, immediately preceding food and in combinations with a constant combination of colours, tones, etc.; while the other stimulus should as regularly be made to act in circumstances which were in all respects different from those just described: say in the late afternoon, out of doors, in a state of considerable activity and exhilaration, nervous excitement, cold and hunger and in combination with an altogether different set of visual and auditory perceptions.

7.48. In using sensory associations to assist the discrimination between stimuli, care would have to be taken not to run counter to well-established synesthetic relations between the qualities of the different senses. The existence of such synesthetic relations between two scales or dimensions of different modalities might, however, well be used to transfer to the other the finer distinctions which the scale of the one modality possesses. Our inadequate knowledge of the character of the stimuli at present probably makes it impossible to use the technique of differential thresholds with regard to olfaction. But as between colours and tones, for instance, persons who have the capacity of colour hearing might

[1]W. J. Brogden, 1939, etc.

well be tested on whether, by deliberately making connexions even closer, the greater capacity of discrimination which they possess in one sense can be transferred to the other.

7.49. It is evident that such a technique for the education of the senses might prove to be of considerable practical importance and should thus be studied even apart from its theoretical significance. It is, of course, more than likely that in such attempts it will be found that the crude approach suggested here is inadequate and that, before much can be accomplished in this direction, much more knowledge about the nature of the sensory order, that is about the interrelations between the dimensions of the various sense modalities, will have to be acquired.

7.50. In addition to such attempts to teach new discriminations between stimuli which were already consciously perceived but not distinguished, the possibility should not be overlooked of attaching conscious values to impulses which did not possess them before. In this connexion perhaps stimuli acting inside the body might offer the most interesting field, and the new techniques of deep heating would seem to open possibilities which ought to be explored. Also the possibility of extending the range of the more familiar senses in this manner should not be disregarded. Although the upper and lower limits of the visible spectrum and the range of audible sounds may well be true physiological limits determined by the nature of the receptor organs, they may in part be centrally determined, and in this case be alterable by training. The considerable inter-individual differences between these limits rather suggest that this may be so, and even such reports as that a blind person has acquired the capacity of smelling colours[1] should not be dismissed as altogether impossible.

7.51. It is not at all improbable that man possesses a considerable number of 'reflex senses', as the action of the semicircular canals in controlling balance has been aptly described,[2] a sensitivity of the body for certain specific stimuli to which a specific response is effected at lower levels, but which have not occurred with sufficient regularity in the company of particular other stimuli to give them a distinct conscious quality. In all such instances it might be

[1] J. T. Williams, 1922, p. 1333. This deserves examination, especially in view of the recent conclusions of Beck and Miles concerning the radiation character of olfactory stimulation.
[2] E. Cyon, quoted by E. G. Boring, 1942, p. 544.

possible to raise these impulses to a conscious level by deliberately attaching to them that characteristic following which they did not have occasion to acquire in natural surroundings.

4. POSSIBILITIES OF EXPERIMENTAL REFUTATION

7.52.   It will assist further to define the content of our main thesis if we state briefly the main alternative theories whose confirmation would at the same time disprove the theory here developed.

7.53.   Disregarding all those theories which, like parallelism, assume the existence of some mind-substance and which are unverifiable almost by definition, the first of the alternative theories which might be mentioned is that of a cell-memory or of the 'storage' of impressions in the individual cell, such as underlies R. Semon's conception of the 'engram'.[1] This conception implies of course, the assumption that whatever it is that is thus stored possesses by itself the different attributes by which different sensory qualities are distinguished. Although it is difficult to see how this assumption could ever be experimentally verified, its confirmation would, of course, refute our theory and in fact eliminate the problem which the latter is intended to solve.

7.54.   A more direct refutation of our theory would be obtained by the discovery of such differences in the physical properties transmitted by the different nerve fibres that these could be said to correspond to the differences in the sensory qualities produced by those impulses—that is, if the theory of the specific energy of nerves in what we have called its illegitimate interpretation (1.33) should prove to be correct. It was by suggesting the search for such physiological differences between the individual impulses that the theoretical views widely held in the past have posed a problem to physiological research to which, if our view is correct, no answer can be found.

7.55.   A special modern form of that theory is the resonance theory developed (for efferent nervous impulses) by P. Weiss[2] which suggests that it is not the fact of a transmission of impulses through special pathways but rather the character of the impulses in some fibres which determines that similar impulses are being set up in other distant fibres. This view, if proved correct for

[1] R. Semon, 1909, 1912.
[2] P. Weiss, 1941.

afferent impulses, would also disprove most of the present theory. The same would be true if the views of some modern gestaltists were confirmed, who seem to suggest that it is not the topological position of the group of impulses in the whole structure of connexions but the spatial configuration[1] of these impulses, irrespective of the particular fibres in which they occur, which counts.

7.56. Finally we might mention as a conceivable alternative theory, although it seems doubtful whether it has ever been carried to its ultimate consequences, the view that sensory discrimination is determined entirely by peripheral motor events. Although we do certainly not wish to minimize the importance of motor responses at all the various levels of the hierarchy of the central nervous system, it is difficult to see how they should ever make those central 'symbolic' or classificatory processes unnecessary with whose functions we were mainly concerned.

[1] W. Köhler and R. Held, 1949.

CHAPTER VIII

PHILOSOPHICAL CONSEQUENCES

1. PRE-SENSORY EXPERIENCE AND PURE EMPIRICISM

8.1.   If the account of the determination of mental qualities which we have given is correct, it would mean that the apparatus by means of which we learn about the external world is itself the product of a kind of experience (5.1–5.16). It is shaped by the conditions prevailing in the environment in which we live, and it represents a kind of generic reproduction of the relations between the elements of this environment which we have experienced in the past; and we interpret any new event in the environment in the light of that experience. If this conclusion is true, it raises necessarily certain important philosophical questions on which in this last chapter we shall attempt some tentative observations.

8.2.   These consequences arise mainly from the rôle which we have assigned to the action of the pre-sensory experience or 'linkages' in determining the sensory qualities. Especially the elimination of the hypothetical 'pure' or 'primary' core of sensations, supposed not to be due to earlier experience, but either to involve some direct communication of properties of the external objects, or to constitute irreducible mental atoms or elements, disposes of various philosophical puzzles which arise from the lack of meaning of those hypotheses.

8.3.   According to the traditional view, experience begins with the reception of sensory data possessing constant qualities which either reflect corresponding attributes belonging to the perceived external objects, or are uniquely correlated with such attributes of the elements of the physical world. These sensory data are supposed to form the raw material which the mind accumulates and learns to arrange in various manners. The theory developed here challenges the basic distinction implied in that conception: the distinction between sensory perception of given qualities and the operations which the intellect is supposed to perform on these data in order to arrive at an understanding of the given phenomenal world (5.19, 6.44).

8.4. According to our theory, the characteristic attributes of the sensory qualities, or the classes into which different events are placed in the process of perception, are not attributes which are possessed by these events and which are in some manner 'communicated' to the mind; they are regarded as consisting entirely in the 'differentiating' responses of the organism by which the qualitative classification or order of these events is created; and it is contended that this classification is based on the connexions created in the nervous system by past linkages. Every sensation, even the 'purest', must therefore be regarded as an interpretation of an event in the light of the past experience of the individual or the species.

8.5. The process of experience thus does not begin with sensations or perceptions, but necessarily precedes them: it operates on physiological events and arranges them into a structure or order which becomes the basis of their 'mental' significance; and the distinction between the sensory qualities, in terms of which alone the conscious mind can learn about anything in the external world, is the result of such pre-sensory experience. We may express this also by stating that experience is not a function of mind or consciousness, but that mind and consciousness are rather products of experience (2.50).

8.6. Every sensory experience of an event in the external world is therefore likely to possess 'attributes' (or to be in a manner distinguished from other sensory events) to which no similar attributes of the external events correspond. These 'attributes' are the significance which the organism has learnt to assign to a class of events on the basis of the past associations of events of this class with certain other classes of events. It is only in so far as the nervous system has learnt thus to treat a particular stimulus as a member of a certain class of events, determined by the connexions which all the corresponding impulses possess with the same impulses representing other classes of events, that an event can be perceived at all, i.e., that it can obtain a distinct position in the system of sensory qualities.

8.7. If the distinctions between the different sensory qualities of which our conscious experience appears to be built up are thus themselves determined by pre-sensory experiences (linkages), the whole problem of the relation between experience and knowledge assumes a new complexion. So far as experience in the narrow sense, i.e., conscious sensory experience, is meant, it is then clearly not true that all that we know is due to such experience. Experience of this kind would rather become possible only after experience in the wider sense of linkages has created the order of sensory qualities—the order which determines the qualities of the constituents of conscious experience.

8.8. Sense experience therefore presupposes the existence of a sort of accumulated 'knowledge', of an acquired order of the sensory impulses based on their past co-occurrence; and this knowledge, although based on (pre-sensory) experience, can never be contradicted by sense experiences and will determine the forms of such experiences which are possible.

8.9. John Locke's famous fundamental maxim of empiricism that *nihil est in intellectu quod non antea fuerit in sensu* is therefore not correct if meant to refer to conscious sense experience. And it does not justify the conclusion that all we know (*quod est in intellectu*) must be subject to confirmation or contradiction by sense experience. From our explanation of the formation of the order of sensory qualities itself it would follow that there will exist certain general principles to which all sensory experiences must conform (such as that two distinct colours cannot be in the same place)—relations between the parts of such experiences which must always be true.

8.10. A certain part at least of what we know at any moment about the external world is therefore not learnt by sensory experience, but is rather implicit in the means through which we can obtain such experience; it is determined by the order of the apparatus of classification which has been built up by pre-sensory linkages. What we experience consciously as qualitative attributes of the external events is determined by relations of which we are not consciously aware but which are implicit in these qualitative distinctions, in the sense that they affect all that we do in response to these experiences.

8.11. All that we can perceive is thus determined by the order of sensory qualities which provides the 'categories' in terms of which sense experience can alone take place. Conscious experience, in particular, always refers to events defined in terms of relations to other events which do not occur in that particular experience.[1]

8.12. We thus possess 'knowledge' about the phenomenal world which, because it is in this manner implicit in all sensory experience, must be true of all that we can experience through our senses. This does not mean, however, that this knowledge must also be true of the physical world, that is, of the order of the stimuli which cause our sensations. While the conditions which make sense perception possible—the apparatus of classification which treats them as similar or different—must affect all sense perception, it does not for this reason also govern the order of the events in the physical world.

8.13. It requires a deliberate effort to divest oneself of the habitual assumption that all we have learned from experience must be true of the external (physical) world.[2] But since all we can ever learn from experience are generalizations about certain kinds of events, and since no number of particular instances can ever prove such a generalization, knowledge based entirely on experience may yet be entirely false. If the significance which a certain group of stimuli has acquired for us is based entirely on the fact that in the past they have regularly occurred in combination with certain other stimuli, this may or may not be an adequate basis for a classification which will enable us to make true predictions. We have earlier (5.20-5.24) given a number of reasons why we must expect that the classifications of events in the external world which our senses perform will not strictly correspond to a classification of these events based solely on the similarity or the differences of their behaviour towards each other.

8.14. While there can thus be nothing in our mind which is not the result of past linkages (even though, perhaps, acquired not by the individual but by the species), the experience that the classification based on the past linkages does not always work, i.e., does not always lead to valid predictions, forces us to revise that classification (6.45-6.48). In the course of this process of

[1] K. Lorenz, 1943, p. 352.

[2] H. von Helmholtz, 1910 (1925), III, p. 14: 'Here we still have to explain how experience counteracts experience, and how illusions can be produced by factors derived from experiences, when it might seem *as if experience could not teach anything except what was true.* In this matter we must remember, as we intimated above, that the sensations are interpreted just as they arise when they are stimulated in the normal way, and when the organ of sense is used normally.' (Italics ours.)

reclassification we not only establish new relations between the data given within a fixed framework of reference, i.e., between the elements of given classes: but since the framework consists of the relations determining the classes, we are led to adjust that framework itself.

8.15. The reclassification, or breaking up of the classes formed by the implicit relations which manifest themselves in our discrimination of sensory qualities, and the replacement of these classes by new classes defined by explicit relations, will occur whenever the expectations resulting from the existing classification are disappointed, or when beliefs so far held are disproved by new experiences. The immediate effects of such conflicting experiences will be to introduce inconsistent elements into the model of the external world; and such inconsistencies can be eliminated only if what formerly were treated as elements of the same class are now treated as elements of different classes (5.72).

8.16. The reclassification which is thus performed by the mind is a process similar to that through which we pass in learning to read aloud a language which is not spelled phonetically. We learn to give identical symbols different values according as they appear in combination with different other symbols, and to recognize different groups of symbols as being equivalent without even noticing the individual symbols.

8.17. While the process of reclassification involves a change of the frame of reference, or of what is *a priori* true of all statements which can be made about the objects defined with respect to that frame of reference, it alters merely the particular presuppositions of all statements, but does not change the fact that such presuppositions must be implied in all statements that can be made. In fact, far from being diminished, the *a priori* element will tend to increase as in the course of this process the various objects are increasingly defined by explicit relations existing between them.

8.18. The new experiences which are the occasion of, and which enter into, the new classifications or definitions of objects, is necessarily presupposed by anything which we can learn about these objects and cannot be contradicted by anything which we can say about the objects thus defined. There is, therefore, on every level, or in every universe of discourse, a part of our knowledge which, although it is the result of experience, cannot be controlled by experience, because it constitutes the ordering principle of that universe by which we distinguish the different kinds of objects of which it consists and to which our statements refer.

8.19. The more this process leads us away from the immediately given sensory qualities, and the more the elements described in terms of these qualities are replaced by new elements defined in terms of consciously experienced relations, the greater becomes the part of our knowledge which is embodied in the definitions of the elements, and which therefore is necessarily true. At the same time the part of our knowledge which is subject to control by experiences becomes correspondingly smaller.

8.20. This progressive growth of the tautological character of our knowledge is a necessary consequence of our endeavour so to readjust our classification of the elements as to make statements about them true. We have no choice but either to accept the classification effected by our senses, and in consequence to be unable correctly to predict the behaviour of the objects thus defined; or to redefine the objects on the basis of the observed differences in their behaviour with respect to each other, with the result that not only the differences which are the basis of our classification become necessarily true of the objects thus classified, but also that it becomes less and less possible to say of any particular sensory object with any degree of certainty to which of our theoretical classes it belongs.

8.21. This difficulty does not become too serious so long as we merely redefine a particular object in relational terms. But as we continue this process of reclassification, those other objects must in turn also be redefined in a similar manner. In the course of this process we are soon forced to take into account not only relations existing between a given object and other objects which are actually observed in conjunction with the former, but also relations which have existed in the past between that and other objects, and even relations which can be described only in hypothetical terms: relations which might have been observed between this and other objects in circumstances which did not in fact exist and which, if they had existed, would not have left the identity of the object unchanged.

8.22. Several chemical substances may, e.g., be completely indistinguishable to the senses so long as they remain in their given state. The reason why chemistry classifies them as different substances is that in certain circumstances and in combination with certain other substances they will 'react' differently. But most of these chemical reactions involve a change in the character of the substance, so that the identical quantity of a given substance, which has been tested for the reaction which is the basis of its classification, cannot be available after it has been established to which class it belongs. Only by such unverifiable assumptions as that the quantity of the substance from which we have drawn the sample is completely homogeneous, so that what we have found out about various samples applies also to the rest, can we arrive at the conclusion that the particular sensory object belongs to a definite theoretical class.

8.23.　The sense data, or the sensory qualities of the objects about which we make statements, thus are pushed steadily further back; and when we complete the process of defining all objects by explicit relations instead of by the implicit relations inherent in our sensory distinctions, those sense data disappear completely from the system. In the end the system of explicit definitions becomes both all-comprehensive and self-contained or circular; all the elements in the universe are defined by their relations to each other, and all we know about that universe becomes contained in those definitions. We should obtain a self-contained model capable of reproducing all the combinations of events which we can observe in the external world, but should have no way of ascertaining whether any particular event in the external world corresponded to a particular part of our model.

8.24.　Science thus tends necessarily towards an ultimate state in which all knowledge is embodied in the definitions of the objects with which it is concerned; and in which all true statements about these objects therefore are analytical or tautological and could not be disproved by any experience. The observation that any object did not behave as it should could then only mean that it was not an object of the kind it was thought to be. With the disappearance of all sensory data from the system, laws (or theories) would no longer exist in it apart from the definitions of the objects to which they applied, and for that reason could never be disproved.

8.25.　Such a completely tautological or self-contained system of knowledge about the world would not be useless. It would constitute a model of the world from which we could read off what kind of events are possible in that world and what kind are not. It would often allow us, on the basis of a fairly complete history of a particular sensory object, to state with a high degree of probability that it fits into one and only one possible place in our model, and that in consequence it is likely to behave in a certain manner in circumstances which would have to be similarly described. But it would never enable us to identify with certainty a particular sensory object with a particular part of our model, or with certainty to predict how the former will behave in given circumstances.

8.26. A strict identification of any point of our theoretical model of the world with a particular occurrence in the sensory world would be possible only if we were in a position to complete our model of the physical world by including in it a complete model of the working of our brain (cf. 5.77–5.91)—that is, if we were able to explain in detail the manner in which our senses classify the stimuli. This, however, as will be shown in section 6 of this chapter, is a task which that same brain can never accomplish.

8.27. In conclusion of this section it should, perhaps, be emphasized that, in so far as we have been led into opposition to some of the theses traditionally associated with empiricism, we have been led to their rejection not from an opposite point of view, but on the contrary, by a more consistent and radical application of its basic idea. Precisely because all our knowledge, including the initial order of our different sensory experiences of the world, is due to experience, it must contain elements which cannot be contradicted by experience. It must always refer to classes of elements which are defined by certain relations to other elements, and it is valid only on the assumption that these relations actually exist. Generalization based on experience must refer to classes of objects or events and can have relevance to the world only in so far as these classes are regarded as given irrespective of the statement itself. Sensory experience presupposes, therefore, an order of experienced objects which precedes that experience and which cannot be contradicted by it, though it is itself due to other, earlier experience.

2. PHENOMENALISM AND THE INCONSTANCY OF SENSORY QUALITIES

8.28.   If the classification of events in the external world effected by our senses proves not to be a 'true' classification, i.e. not one which enables us adequately to describe the regularities in this world, and if the properties which our senses attribute to these events are not objective properties of these individual events, but merely attributes defining the classes to which our senses assign them, this means that we cannot regard the phenomenal world in any sense as more 'real' than the constructions of science: we must assume the existence of an objective world (or better, of an objective order of the events which we experience in their phenomenal order) towards the recognition of which the phenomenal order is merely a first approximation. The task of science is thus to try and approach ever more closely towards a reproduction of this objective order—a task which it can perform only by replacing the sensory order of events by a new and different classification.[1]

8.29.   By saying that there 'exists' an 'objective' world different from the phenomenal world we are merely stating that it is possible to construct an order or classification of events which is different from that which our senses show us and which enables us to give a more consistent account of the behaviour of the different events in that world. Or, in other words, it means that our knowledge of the phenomenal world raises problems which can be answered only by altering the picture which our senses give us of that world, i.e. by altering our classification of the elements of which it consists. That this is possible and necessary is, in fact, a postulate which underlies all our efforts to arrive at a scientific explanation of the world.

8.30.   Any purely phenomenalistic interpretation of the task of science, or any attempt to reduce this task to merely a complete description of the phenomenal world, thus must break down because our senses do not effect such a classification of the different events that what appears to us as alike will also always behave in the same manner. The basic thesis of phenomenalism (and positivism) that 'all *phenomena* are subject to invariable laws' is simply not true if the term phenomenon is taken in its strict meaning of things as they appear to us.

[1]Cf., M. Planck, 1942 (1949), p. 108.

8.31. The ideal of science as merely a complete description of phenomena, which is the positivist conclusion derived from the phenomenalistic approach, therefore proves to be impossible. Science consists rather in a constant search for new classes, for 'constructs' which are so defined that general propositions about the behaviour of their elements are universally and necessarily true. For this purpose these classes cannot be defined in terms of sensory properties of the particular individual events perceived by the individual person; they must be defined in terms of their relations to other individual events.

8.32. Such a definition of any class of events, in terms of their relations to other classes of events instead of in terms of any sensory properties which they individually possess, cannot be confined to the former, or even to all the events which together constitute the complete situation existing at a particular moment. The events referred to in the definition of those with which we have actually to deal have to be defined in a similar purely 'relational' manner. The ultimate aim of this procedure must be to define all classes of events exclusively in terms of their relations to each other and without any reference to their sensory properties. It has been well said that 'for science an object expresses itself in the totality of relations possible between it and other objects.'[1] We have already seen (8.24–8.25) that such a complete system of explanation would necessarily be tautological, because all that could be predicted by it would necessarily follow from the definitions of the objects to which it referred.

8.33. If the theory outlined here is correct, there exists an even more fundamental objection to any consistently phenomenalist interpretation of science. It would appear that not only are the events of the world, if defined in terms of their sensory attributes, not subject to invariable laws, so that situations presenting the same appearance to our senses may produce different results; but also that the phenomenal world (or the order of the sensory qualities from which it is built up) is itself not constant but variable, and that it will in some measure change its appearance as a result of that very process of reclassification which we must perform in order to explain it.

8.34. If it is true, as we have argued, that the 'higher' mental activities are merely a repetition at successive levels of processes

[1] *Fundamental Mathematics*, 1948, I, p. 92.

of classification of essentially the same character as those by which the different sensory qualities have come to be distinguished in the first instance, it would seem almost inevitable that this process of reclassification will in some measure also affect the distinctions between the different sensory qualities from which it starts. The nature of the process by which the difference between sensory qualities are determined makes it probable that they will remain variable and that the distinctions between them will be modified by new experiences. This would mean that the phenomenal world itself would not be constant but would be incessantly changing in a direction to a closer reproduction of the relations existing in the physical world. If in the course of this process the sensory data themselves alter their character, the ideal of a purely descriptive science becomes altogether impossible.

8.35. That the sensory qualities which attach to particular physical events are thus in principle themselves variable[1] is no less important even though we must probably regard them as *relatively* stable compared with the continuous changes of the scheme of classification in terms of which abstract thought proceeds, almost certainly in so far as the course of the life of the individual is concerned. But we should still have to consider more seriously than we are wont to do, what is amply confirmed by ordinary experience, namely that as a result of the advance of our explanation of the world we also come to 'see' this world differently, i.e. that we not merely recognize new laws which connect the given phenomena, but that these events are themselves likely to change their appearance to us.

8.36. Such variations of the sensory qualities attributed to given events could, of course, never be ascertained by direct comparisons of past and present sensations, since the memory images of past sensations would be subject to the same changes as the current sensations. The only possibility of testing this conclusion would be provided by experiments with discrimination such as have been suggested in the preceding chapter (7.38–7.51).

8.37. It deserves, perhaps, to be mentioned that, although the

---

[1]This changeability of the sensory qualities apparently was already recognized by Protagoras, who according to Sextus Empiricus taught that the sensations 'are transformed and altered according to the times of life and to all the other conditions of the body.' *Outlines of Pyrrhonianism*, translation R. G. Bury, *Loeb Classical Library*, I, Book I, 218.

theory developed here was suggested in the first instance by the psychological views which Ernst Mach has outlined in his *Analysis of Sensations* and elsewhere, its systematic development leads to a refutation of his and similar phenomenalist philosophies: by destroying the conception of elementary and constant sensations as ultimate constituents of the world, it restores the necessity of a belief in an objective physical world which is different from that presented to us by our senses.[1]

8.38. Similar considerations apply to the views expounded on these matters by William James, John Dewey and the American realists and developed by Bertrand Russell. The latter's view according to which 'the stuff of the world' consists of 'innumerable transient particulars' such as a patch of colour which is 'both physical and psychical' in fact is explicitly based on the assumption that 'sensations are what is common to the mental and the physical world', and that their essence is 'their independence from past experience'. The whole of this 'neutral monism' seems to be based on entirely untenable psychological conceptions.[2]

8.39. Another interesting consequence following from our theory is that a stimulus whose occurrence in conjunction with other stimuli showed no regularities whatever could never be perceived by our senses (6.36). This would seem to mean that we can know only such kinds of events as show a certain degree of regularity in their occurrence in relations with others, and that we could not know anything about events which occurred in a completely irregular manner. The fact that the world which we know seems wholly an orderly world may thus be merely a result of the method by which we perceive it. Everything which we can perceive we perceive necessarily as an element of a class of events which obey certain regularities. There could be in this sense no class of events showing no regularities, because there would be nothing which could constitute them for us into a distinct class.

---

[1] Cf., K. Koffka, 1935, p. 63: 'Mach was an excellent psychologist, who saw many of the most fundamental problems of psychology which, a whole generation later, many psychologists failed even to understand; at the same time he had a philosophy which made it impossible to give fruitful solutions to these problems.'

[2] B. Russell, 1921, p. 144.

3. DUALISM AND MATERIALISM

8.40.   Because the account of the determination of mental qualities which has been given here explains them by the operation of processes of the same kind as those which we observe in the material world, it is likely to be described as a 'materialistic' theory. Such a description in itself would matter very little if it were not for certain erroneous ideas associated with the term materialism which not only would prejudice some people against a theory thus described but, what is more important, would also suggest that it implies certain conclusions which are almost the opposite of those which in fact follow from it. In the true sense of the word 'materialistic' it might even be argued that our theory is less materialistic than the dualistic theories which postulate a distinct mind 'substance'.

8.41.   The dualistic theories are a product of the habit, which man has acquired in his early study of nature, of assuming that in every instance where he observed a peculiar and distinct process it must be due to the presence of a corresponding peculiar and distinct substance. The recognition of such a peculiar material substance came to be regarded as an adequate explanation of the process produced.

8.42.   It is a curious fact that, although in the realm of nature in general we no longer accept as an adequate explanation the postulate of a peculiar substance possessing the capacity of producing the phenomena we wish to explain, we still resort to this old habit where mental events are concerned. The mind 'stuff' or 'substance' is a conception formed in analogy to the different kinds of matter supposedly responsible for the different kinds of material phenomena. It is, to use an old term in its literal sense, the result of a 'hylomorphic' manner of thinking. Yet in whatever manner we define substance, to think of mind as a substance is to ascribe to mental events some attributes for whose existence we have no evidence and which we postulate solely on the analogy of what we know of material phenomena.[1]

8.43.   In the strict sense of the terms employed an account of mental phenomena which avoids the conception of a distinct mental substance is therefore the opposite of materialistic, because it does not attribute to mind any property which we derive from

[1]This seems to me to be true in spite of the efforts of C. D. Broad, 1925, to give 'substance' a meaning independent of its material connotations. On the mind substance theory see now G. Ryle, 1949.

our acquaintance with matter. In being content to regard mind as a peculiar order of events, different from the order of events which we encounter in the physical world, but determined by the same kind of forces as those that rule in that world, it is indeed the only theory which is not materialistic.[1]

8.44. Superficially there may seem to exist a closer connexion between the theory presented here and the so-called 'double aspect theories' of the relations between mind and body. To scribe our theory as such, however, would be misleading. What could be regarded as the 'physical aspect' of this double-faced entity would not be the individual neural processes but only the complete order of all these processes; but this order, if we knew it in full, would then not be another aspect of what we know as mind but would be mind itself. We cannot directly observe how this order is formed by its physical elements, but can only infer it. But if we could complete the theoretical reconstruction of this order from its elements and then disregard all the properties of the elements which are not relevant to the existence of this order as a whole, we should have a complete description of the order we call mind—just as in describing a machine we can disregard many properties of its parts, such as their colour, and consider only those which are essential to the functioning of the machine as a whole. (Cf. 2.28–2.30.)

8.45. This order which we call mind is thus the order prevailing in a particular part of the physical universe—that part of it which is ourselves. It is an order which we 'know' in a way which is different from the manner in which we know the order of the physical universe around us. What we have tried to do here is to show that the same kind of regularities which we have learnt to discover in the world around us are in princple also capable of building up an order like that constituting our mind. That such a kind of sub-order can be formed within that order which we have discovered in the external universe does not yet mean, however, that we must be able to explain how the particular order which constitutes our mind is placed in that more comprehensive order. In order to achieve this it would be necessary to construct, with special

---

[1]Cf., *N*. Metzger, 1941, p. 23: 'Diese, im eigentlich Sinn "materialistische" Auffassung . . . lebt in der Psychologie bis an die Schwelle unserer Zeit fort: in der Alltagspsychologie in der kaum ausrottbaren Ansicht von der Seele als zweitem, stofflichen Etwas, das mit dem Körper während des Lebens "verbunden sei", in ihm wohne . . .'

reference to the human mind, a detailed reproduction of the model-object relation which it involves such as we have sketched schematically before in order to illustrate the general principle (5.77–5.91).

8.46. While our theory leads us to deny any ultimate dualism of the forces governing the realms of mind and that of the physical world respectively, it forces us at the same time to recognize that for practical purposes we shall always have to adopt a dualistic view. It does this by showing that any explanation of mental phenomena which we can hope ever to attain cannot be sufficient to 'unify' all our knowledge, in the sense that we should become able to substitute statements about particular physical events (or classes of physical events) for statements about mental events without thereby changing the meaning of the statement.

8.47. In this specific sense we shall never be able to bridge the gap between physical and mental phenomena; and for practical purposes, including in this the procedure appropriate to the different sciences, we shall permanently have to be content with a dualistic view of the world. This, however, raises a further problem which must be more systematically considered in the remaining sections of this chapter.

4. THE NATURE OF EXPLANATION

8.48. What remains now is to restate briefly what the theory outlined in the preceding pages is meant to explain, and how far it can be expected to account for particular mental processes. This makes it necessary to make more precise than we have yet done what we mean by 'explanation'. This is a peculiarly relevant question since 'explanation' is itself one of the mental processes which the theory intends to explain.

8.49. It has been suggested before (5.44–5.48) that explanation consists in the formation in the brain of a 'model' of the complex of events to be explained, a model the parts of which are defined by their position in a more comprehensive structure of relationships which constitutes the semi-permanent framework from which the representations of individual events receive their meaning.

8.50. This notion of the 'model', which the brain is assumed to be capable of building, has, of course, been often used in this connexion,[1] and by itself it does not get us very far. Indeed, if it is

[1]See particularly K. J. W. Craik, 1943, and K. Lorenz, 1943, pp. 343 and 351.

conceived, as is usually the case, as a separate model of the particular phenomenon to be explained, it is not at all clear what is meant by it. The analogy with a mechanical model is not directly applicable. A mechanical model derives its significance from the fact that the properties of its individual parts are assumed to be known and in some respects to correspond to the properties of the parts of the phenomenon which it reproduces. It is from this knowledge of the different properties of the parts that we derive our knowledge of how the particular combination of these parts will function.

8.51. In general, the possibility of forming a model which explains anything presupposes that we have at our disposal distinct elements whose action in different circumstances is known irrespective of the particular model in which we use them. In the case of a mechanical model it is the physical properties of the individual parts which are supposed to be known. In a mathematical 'model' the 'properties' of the parts are defined by functions which show the values they will assume in different circumstances, and which are capable of being combined into various systems of equations which constitute the models.

8.52. The weakness of the ordinary use of the concept of the model as an account of the process of explanation consists in the fact that this conception presupposes, but does not explain, the existence of the different mental entities from which such a model could be built. It does not explain in what sense or in which manner the parts of the model correspond to the parts of the original, or what are the properties of the elements from which the model is built.

8.53. The concept of a model that is being formed in the brain is helpful only after we have succeeded in accounting for the different properties of the parts from which it is built. Such an account is provided by the explanation of the determination of sensory (and other mental) qualities by their position in the more comprehensive semi-permanent structure of relationships, the 'map' of the world which past experience has created in the brain, which has been described in the preceding pages. It is the position of the impulse in the connected network of fibres which brings it about that its occurrence together with other impulses will produce certain further impulses. The formation of the model appears thus merely a particular case of that process of joint or simultaneous

classification of a group of impulses of which each has its determined significance apart from the particular combination or
model in which it now occurs.

8.54. We can schematically represent this process of joint classification which produces a model in the following manner: the
different elements, the mental qualities from which the model is
built, are classes of impulses which we may call A, B, C, etc., and
which are defined as an a (member of A) producing x (and perhaps some other impulses) when it occurs in company with
o, p, . . ., but producing v, z, . . . when it occurs in company with
r, s, . . . etc., etc., and similarly for all members of the classes
B, C, etc. In this definition any given class of impulses may, of
course, occur both in a 'primary' character, that is as an element
of a class to be defined by the impulses which any element of this
class will evoke, and in a 'secondary' character as an evoked
impulse which determines the class to which some other impulses
belong. (3.38, 3.55ff.) Impulses of the class A will appear not only
in statements like 'if (a, o, p) then x' and 'if (a, r, s) then$(y, z \ldots)$,
but also in statements like 'if (b, c, q) then (a, t) etc.

8.55. Given such a determination of the different significance of
impulses of the different classes, it follows that any given combination of such impulses will produce impulses standing for other
classes, and these in turn others, and so on, somewhat as in the
following schematic representation:

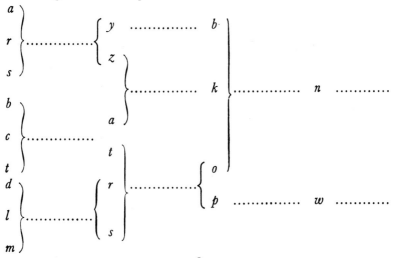

8.56. The particular result produced is thus recognized to be the effect of the simultaneous occurrence of certain elements in a particular constellation which, if we had known of their presence, would have enabled us to predict the result. Once we have formed such a model we are in a position to say on which of the various elements in the actual situation the observed result depends, and how it would be modified if any of these elements were changed; this is what an explanation enables us to do.

5. EXPLANATION OF THE PRINCIPLE

8.57. It follows from what has been said so far about explanation that it will always refer to classes of events, and that it will account only for those properties which are common to the elements of the class. Explanation is always generic in the sense that it always refers to features which are common to all phenomena of a certain kind, and it can never explain everything to be observed on a particular set of events.

8.58. But although all explanation must refer to the common features of a class of phenomena, there are evidently different degrees to which an explanation can be general, or to which it may approach to a full explanation of a particular set of events. The model may reproduce only the few common features of a great variety of phenomena, or it may reproduce a much larger number of features common to a smaller number of instances. In general, it will be true that the simpler the model, the wider will be the range of particular phenomena of which it reproduces one aspect, and the more complex the model, the more will its range of application be restricted.[1] In this respect the relation of the model to the object is similar to that between the connotation and the denotation (or the 'intension' and the 'extension') of a concept.

8.59. Most explanations (or theories) with which we are familiar are intended to show a common principle which operates in a large number of particular instances which in other respects may differ widely from each other. We have referred already earlier (2.18–2.19) to such explanations as 'explanations of the principle'.[2] The difference between such 'explanations of the principle' and more detailed explanations is, of course, merely one of the degree

[1] Cf., M. Petrovitch, 1921, passim.
[2] See also F. A. Hayek, 1942, p. 290.

of their generality, and strictly speaking no explanation can be more than an explanation of the principle. It will be convenient, however, to reserve the name 'explanation of the principle' for explanations of a high degree of generality, and to contrast them with explanations of the detail.

8.60. The usual kind of explanation which we give, e.g., of the functioning of a clockwork, will in our sense be merely an explanation of the principle. It will merely show how the kind of phenomena which we call clockworks are produced: the manner in which a pair of hands can be made to revolve at constant speeds, etc. In the same 'general' way most of us are familiar with the principles on which a steam engine, an atomic bomb, or certain kinds of simple organisms function, without therefore necessarily being able to give a sufficiently detailed explanation of any one of these objects so that we should be able to construct it or precisely to predict its behaviour. Even where we are able to construct one of these objects, say a clockwork, the knowledge of the principle involved will not be sufficient to predict more than certain general aspects of its operation. We should never be able, for instance, before we have built it, to predict precisely how fast it will move or precisely where its hands will be at a particular moment of time.

8.61. If in general we are not more aware of this distinction between explanations merely of the principle and more detailed explanations, this is because usually there will be no great difficulty about elaborating any explanation of the principle so as to make it approximate to almost any desired degree to the circumstances of a particular situation. By increasing the complexity of the model we can usually obtain a close reproduction of any particular feature in which we are interested.

8.62. The distinction between the explanation of the principle on which a wide class of phenomena operate and the more detailed explanation of particular phenomena is reflected in the familiar distinction between the 'theoretical' and the more 'applied' parts of the different sciences. 'Theoretical physics', 'theoretical chemistry' or 'theoretical biology' are concerned with the explanation of the principles common to all phenomena which we call physical, chemical or biological.

8.63. Strictly speaking we should, of course, not be entitled to speak at all of phenomena of a certain kind unless we know some

common principles which apply to the explanation of the phenomena of that kind. The various ways in which atoms are combined
into molecules, e.g., constitute the common principles of all the
phenomena which we call chemical. It is quite possible that an
observed phenomenon, supposed to be, say, chemical, such as a
change in the colour of a certain substance, may on investigation
prove to be an event of a different kind, e.g., an optical event, such
as a change in the light falling on the substance.

8.64. While it is true that a theoretical class of phenomena can be
definitely established only after we have found a common principle of explanation applying to all its members, that is, a model
of high degree of generality reproducing the features they all have
in common, we will yet often know of a range of phenomena which
seem to be similar in some respect and where we therefore expect
to find a common principle of explanation without, however, as
yet knowing such a principle. The difference between such *prima
facie* or 'empirical' classes of phenomena, and the theoretical
classes derived from a common principle of explanation, is that
the empirical class is limited to phenomena actually observed,
while the theoretical class enables us to define the range over
which phenomena of the kind in question may vary.

8.65. The class of events which we call 'mental' has so far on the
whole been an empirical class in this sense. What has been
attempted here might be described as a sketch of a 'theoretical
psychology' in the same sense in which we speak of theoretical
physics or theoretical biology. We have attempted an explanation
of the principle by which we may account for the peculiarities
which are common to all processes which are commonly called
mental. The question which arises now is how far in the sphere of
mental processes we can hope to develop the explanation of the
principle into more detailed explanations, especially into explanations that would enable us to predict the course of particular
mental events.

6. THE LIMITS OF EXPLANATION

8.66. It is by no means always and necessarily true that the
achievement of an explanation of the principle on which the
phenomena of a certain class operate enables us to proceed to
explanations of the more concrete detail. There are several fields

in which practical difficulties prevent us from thus elaborating known explanations of the principle to the point where they would enable us to predict particular events. This is often the case when the phenomena are very complex, as in meteorology or biology; in these instances, the number of variables which would have to be taken into account is greater than that which can be ascertained or effectively manipulated by the human mind. While we may, e.g., possess full theoretical knowledge of the mechanism by which waves are formed and propagated on the surface of water, we shall probably never be able to predict the shape and movements of the wave that will form on the ocean at a particular place and moment of time.

8.67.   Apart from these practical limits to explanation, which we may hope continuously to push further back, there also exists, however, an absolute limit to what the human brain can ever accomplish by way of explanation—a limit which is determined by the nature of the instrument of explanation itself, and which is particularly relevant to any attempt to explain particular mental processes.

8.68.   If our account of the process of explanation is correct, it would appear that any apparatus or organism which is to perform such operations must possess certain properties determined by the properties of the events which it is to explain. If explanation involves that kind of joint classification of many elements which we have described as 'model-building', the relation between the explaining agent and the explained object must satisfy such formal relations as must exist between any apparatus of classification and the individual objects which it classifies. (Cf. 5.77–5.91.)

8.69.   The proposition which we shall attempt to establish is that any apparatus of classification must possess a structure of a higher degree of complexity than is possessed by the objects which it classifies; and that, therefore, the capacity of any explaining agent must be limited to objects with a structure possessing a degree of complexity lower than its own. If this is correct, it means that no explaining agent can ever explain objects of its own kind, or of its own degree of complexity, and, therefore, that the human brain can never fully explain its own operations. This statement possesses, probably, a high degree of *prima facie* plausibility. It is, however, of such importance and far-reaching consequences, that we must attempt a stricter proof.

8.70. We shall attempt such a demonstration at first for the simple processes of classification of individual elements, and later apply the same reasoning to those processes of joint classification which we have called model-building. Our first task must be to make clear what we mean when we speak of the 'degree of complexity' of the objects of classification and of the classifying apparatus. What we require is a measure of this degree of complexity which can be expressed in numerical terms.

8.71. So far as the objects of classification are concerned, it is necessary in the first instance to remember that for our purposes we are not interested in all the properties which a physical object may possess in an objective sense, but only in those 'properties' according to which these objects are to be classified. For our purposes the complete classification of the object is its complete definition, containing all that with which we are concerned in respect to it.

8.72. The degree of complexity of the objects of classification may then be measured by the number of different classes under which it is subsumed, or the number of different 'heads' under which it is classified. This number expresses the maximum number of points with regard to which the response of the classifying apparatus to this object may differ from its responses to any one other object which it is also capable of classifying. If the object in question is classified under n heads, it can evidently differ from any one other object that is classified by the same apparatus in n different ways.

8.73. In order that the classifying apparatus should be able to respond differently to any two objects which are classified differently under any one of these n heads, this apparatus will clearly have to be capable of distinguishing between a number of classes much larger than n. If any individual object may or may not belong to any one of the n classes $A, B, C, \ldots N$, and if all individual objects differing from each other in their membership of any one of these classes are to be treated as members of separate classes, then the number of different classes of objects to which the classifying apparatus will have to be able to respond differently will, according to a simple theorem of combinatorial analysis, have to be 2^{n+1}.

8.74. The number of different responses (or groups of responses), of which the classifying apparatus is capable, or the number of different classes it is able to form, will thus have to be of a definitely higher order of magnitude than the number of classes to which any individual object of classification can belong. This remains true when many of the individual classes to which a particular object belongs are mutually exclusive or disjunct, so that it can belong only to either A_1 or A_2 or A_3 ... and to either B_1 or B_2 or B_3 ... · etc. If in such a case the number of variable 'attributes' which distinguish elements of A_1 from elements of A_2 and of A_3, and elements of B_1 from those of B_2 and B_3, etc., is m and each of these m different variable attributes may assume n different 'values', although any one element will belong to at most m different classes, the number of distinct combinations of attributes to which the classifying apparatus will have to respond will still be equal to n^m.

8.75. In the same way in which we have used the number of different classes to which any one element can be assigned as the measure of the degree of its complexity, we can use the number of different classes to which the classifying apparatus will have to respond differently as the measure of complexity of that apparatus. It is evidently this number which indicates the variety of ways in which any one scheme of classification for a given set of elements may differ from any other scheme of classification for the different schemes of classification which can be applied to the given set of elements. Such a scheme for classifying the different possible schemes of classification would in turn have to possess a degree of complexity as much greater than that of any of the latter as their degree of complexity exceeds the complexity of any one of the elements.

8.76. What is true of the relationship between the degree of complexity of the different elements to be classified and that of the apparatus which can perform such classification, is, of course, equally true of that kind of joint or simultaneous classification which we have called 'model-building'. It differs from the classification of individual elements merely by the fact that the range of possible differences between different constellations of such elements is already of a higher order of magnitude than the range of possible differences between the individual elements, and that in consequence any apparatus capable of building models of all the different possible constellations of such elements must be of an even higher order of complexity.

8.77. An apparatus capable of building within itself models of different constellations of elements must be more complex, in our sense, than any particular constellation of such elements of which it can form a model, because, in addition to showing how any one of these elements will behave in a particular situation, it must be capable also of representing how any one of these elements would behave in any one of a large number of other situations. The 'new' result of the particular combination of elements which it is capable of predicting is derived from its capacity of predicting the behaviour of each element under varying conditions.

8.78. The significance of these abstract considerations will become clearer if we consider as illustrations some instances in which this or a similar principle applies. The simplest illustration of the kind is probably provided by a machine designed to sort out certain objects according to some variable property. Such a machine will clearly have to be capable of indicating (or of differentially responding to) a greater number of different properties than any one of the objects to be sorted will possess. If, e.g., it is designed to sort out objects according to their length, any one object can possess only one length, while the machine must be capable of a different response to many different lengths.

8.79. An analogous relationship, which makes it impossible to work out on any calculating machine the (finite) number of distinct operations which can be performed with it, exists between that number and the highest result which the machine can show. If that limit were, e.g., 999,999,999, there will already be 500,000,000 additions of two different figures giving 999,999,999 as a result, 499,999,999 pairs of different figures the addition of which gives 999,999,998 as a result, etc. etc., and therefore a very much larger number of different additions of pairs of figures only than the machine can show. To this would have to be added all additions of more than two figures and all the different instances of the other operations which the machine can perform. The number of distinct calculations it can perform therefore will be clearly of a higher order of magnitude than the highest figure it can enumerate.

8.80. Applying the same general principle to the human brain as an apparatus of classification it would appear to mean that, even though we may understand its *modus operandi* in general terms, or, in other words, possess an explanation of the principle on which it operates, we shall never, by means of the same brain, be able to arrive at a detailed explanation of its working in particular circumstances, or be able to predict what the results of its operations will be. To achieve this would require a brain of a higher order of complexity, though it might be built on the same general principles. Such a brain might be able to explain what happens in our brain, but it would in turn still be unable fully to explain its own operations, and so on.

8.81. The impossibility of explaining the functioning of the human brain in sufficient detail to enable us to substitute a description in physical terms for a description in terms of mental qualities, applies thus only in so far as the human brain is itself to be used as the instrument of classification. It would not only not apply to a brain built on the same principle but possessing a higher order of complexity, but, paradoxical as this may sound, it also does not exclude the logical possibility that the knowledge of the principle on which the brain operates might enable us to build a machine fully reproducing the action of the brain and capable of predicting how the brain will act in different circumstances.

8.82. Such a machine, designed by the human mind yet capable of 'explaining' what the mind is incapable of explaining without its help, is not a self-contradictory conception in the sense in which the idea of the mind directly explaining its own operations involves a contradiction. The achievement of constructing such a machine would not differ in principle from that of constructing a calculating machine which enables us to solve problems which have not been solved before, and the results of whose operations we cannot, strictly speaking, predict beyond saying that they will be in accord with the principles built into the machine. In both instances our knowledge merely of the principle on which the machine operates will enable us to bring about results of which, before the machine produces them, we know only that they will satisfy certain conditions.

8.83. It might appear at first as if this impossibility of a full explanation of mental processes would apply only to the mind as a whole, and not to particular mental processes, a full explanation of which might still enable us to substitute for the description of a particular mental process a fully equivalent statement about a set of physical events. Such a complete explanation of any particular mental process would, if it were possible, of course be something different from, and something more far-reaching than, the kind of partial explanation which we have called 'explanation of the principle.'

8.84. In order to provide a full explanation of even one particular mental process, such an explanation would have to run entirely in physical terms, and would have to contain no references to any other mental events which were not also at the same time explained in physical terms. Such a possibility is ruled out, however, by the fact, that the mind as an order is a 'whole' in the strict sense of the term: the distinct character of mental entities and of their mode of operation is determined by their relations to (or their position in the system of) all other mental entities. No one of them can, therefore, be explained without at the same time explaining all the others, or the whole structure of relationships determining their character.

8.85. So long as we cannot explain the mind as a whole, any attempt to explain particular mental processes must therefore contain references to other mental processes and will thus not achieve a full reduction to a description in physical terms. A full translation of the description of any set of events from the mental to the physical language would thus presuppose knowledge of the complete set of 'rules of correspondence'[1] by which the two languages are related, or a complete account of the orders prevailing in the two worlds.

8.86. This conclusion may be expressed differently by saying that a mental process could be identified with (or 'reduced to') a particular physical process only if we were able to show that it occupies in the whole order of mental events a position which is identical with the position which the physical events occupy in the physical order of the organism. The former is a mental process because it occupies a certain position in the whole order of mental process (i.e., because of the manner in which it can affect, and be affected by, other mental processes), and this position in an order can be explained in physical terms only by showing how an equivalent order can be built up from physical elements. Only if we could achieve this could we substitute for our knowledge of mental events a statement of the order existing in a particular part of the physical world.

[1] H. Margenau, 1950, pp. 60, 69, 450.

7. THE DIVISION OF THE SCIENCES AND THE 'FREEDOM OF THE WILL'

8.87. The conclusion to which our theory leads is thus that to us not only mind as a whole but also all individual mental processes must forever remain phenomena of a special kind which, although produced by the same principles which we know to operate in the physical world, we shall never be able fully to explain in terms of physical laws. Those whom it pleases may express this by saying that in some ultimate sense mental phenomena are 'nothing but' physical processes; this, however, does not alter the fact that in discussing mental processes we will never be able to dispense with the use of mental terms, and that we shall have permanently to be content with a practical dualism, a dualism based not on any assertion of an objective difference between the two classes of events, but on the demonstrable limitations of the powers of our own mind fully to comprehend the unitary order to which they belong.

8.88. From the fact that we shall never be able to achieve more than an 'explanation of the principle' by which the order of mental events is determined, it also follows that we shall never achieve a complete 'unification' of all sciences in the sense that all phenomena of which it treats can be described in physical terms.[1] In the study of human action, in particular, our starting point will always have to be our direct knowledge of the different kinds of mental events, which to us must remain irreducible entities.

8.89. The permanent cleavage between our knowledge of the physical world and our knowledge of mental events goes right through what is commonly regarded as the one subject of psychology. Since the theoretical psychology which has been sketched here can never be developed to the point at which it would enable us to substitute for the description of particular mental events descriptions in terms of particular physical events, and since it has

---

[1] The term 'physical' must here be understood in the strict sense in which it has been defined in the first chapter and not be confused with the sense in which it is used, e.g., by O. Neurath or R. Carnap when they speak of the 'physical language'. In our sense their 'physical language', since it refers to the phenomenal or sensory qualities of the objects, is not 'physical' at all. Their use of this term rather implies a metaphysical belief in the ultimate 'reality' and constancy of the phenomenal world for which there is little justification. Cf., O. Neurath, 1933, and R. Carnap, 1934.

therefore nothing to say about particular kinds of mental events, but is confined to describing the *kind* of physical processes by which the various types of mental processes can be produced, any discussion of mental events which is to get beyond such a mere 'explanation of the principle' will have to start from the mental entities which we know from direct experience.

8.90. This does not mean that we may not be able in a different sense to 'explain' particular mental events: it merely means that the type of explanation at which we aim in the physical sciences is not applicable to mental events. We can still use our direct ('introspective') knowledge of mental events in order to 'understand,' and in some measure even to predict, the results to which mental processes will lead in certain conditions. But this introspective psychology, the part of psychology which lies on the other side of the great cleavage which divides it from the physical sciences, will always have to take our direct knowledge of the human mind for its starting point. It will derive its statements about some mental processes from its knowledge about other mental processes, but it will never be able to bridge the gap between the realm of the mental and the realm of the physical.

8.91. Such a *verstehende* psychology, which starts from our given knowledge of mental processes, will, however, never be able to explain why we must think thus and not otherwise, why we arrive at particular conclusions. Such an explanation would presuppose a knowledge of the physical conditions under which we would arrive at different conclusions. The assertion that we can explain our own knowledge involves also the belief that we can at any one moment of time both act on some knowledge and possess some additional knowledge about how the former is conditioned and determined. The whole idea of the mind explaining itself is a logical contradiction—nonsense in the literal meaning of the word—and a result of the prejudice that we must be able to deal with mental events in the same manner as we deal with physical events.[1]

8.92. In particular, it would appear that the whole aim of the discipline known under the name of 'sociology of knowledge' which aims at explaining why people as a result of particular material circumstances hold particular views at particular

[1]On this and the subject of the next paragraph cf., F. A. Hayek, 1944, pp. 31 *et seq.*

moments, is fundamentally misconceived. It aims at precisely that kind of specific explanation of mental phenomena from physical facts which we have tried to show to be impossible. All we can hope to do in this field is to aim at an explanation of the principle such as is attempted by the general theory of knowledge or epistemology.

8.93. It may be noted in passing that these considerations also have some bearing on the age-old controversy about the 'freedom of the will'. Even though we may know the general principle by which all human action is causally determined by physical processes, this would not mean that to us a particular human action can ever be recognizable as the necessary result of a particular set of physical circumstances. To us human decisions must always appear as the result of the whole of a human personality—that means the whole of a person's mind—which, as we have seen, we cannot reduce to something else.[1]

8.94. The recognition of the fact that for our understanding of human action familiar mental entities must always remain the last determinants to which we can penetrate, and that we cannot hope to replace them by physical facts, is, of course, of the greatest importance for all the disciplines which aim at an understanding and interpretation of human action. It means, in particular, that the devices developed by the natural sciences for the special purpose of replacing a description of the world in sensory or phenomenal terms by one in physical terms lose their *raison d'être* in the study of intelligible human action. This applies particularly to the endeavour to replace all qualitative statements by quantitative expressions or by descriptions which run exclusively in terms of explicit relations.[2]

8.95. The impossibility of any complete 'unification' of all our scientific knowledge into an all-comprehensive physical science has hardly less significance, however, for our understanding of the physical world than it has for our study of the consequences of human action. We have seen how in the physical sciences the aim

[1]It may also be mentioned, although this has little immediate connexion with our main subject, that since the word 'free' has been formed to describe a certain subjective experience and can scarcely be defined except by reference to that experience, it could at most be asserted that the term is meaningless. But this would make any denial of the existence of free will as meaningless as its assertion.

[2]For a fuller discussion of this point see F. A. Hayek, 1942, p. 290 ff.

is to build models of the connexions of the events in the external world by breaking up the classes known to us as sensory qualities and by replacing them by classes explicitly defined by the relations of the events to each other; also how, as this model of the physical world becomes more and more perfect, its application to any particular phenomenon in the sensory world becomes more and more uncertain. (8.17–8.26.)

8.96. A definite co-ordination of the model of the physical world thus constructed with the picture of the phenomenal world which our senses give us would require that we should be able to complete the task of the physical sciences by an operation which is the converse of their characteristic procedure (1.21): we should have to be able to show in what manner the different parts of our model of the physical world will be classified by our mind. In other words, a complete explanation of even the external world as we know it would presuppose a complete explanation of the working of our senses and our mind. If the latter is impossible, we shall also be unable to provide a full explanation of the phenomenal world.

8.97. Such a completion of the task of science, which would place us in a position to explain in detail the manner in which our sensory picture of the external world represents relations existing between the parts of this world, would mean that this reproduction of the world would have to include a reproduction of that reproduction (or a model of the model-object relation) which would have to include a reproduction of that reproduction of that reproduction, and so on *ad infinitum*. The impossibility of fully explaining any picture which our mind forms of the external world therefore also means that it is impossible ever fully to explain the 'phenomenal' external world. The very conception of such a completion of the task of science is a contradiction in terms. The quest of science is, therefore, by its nature a never-ending task in which every step ahead with necessity creates new problems.

8.98. Our conclusion, therefore, must be that *to us* mind must remain forever a realm of its own which we can know only through directly experiencing it, but which we shall never be able fully to explain or to 'reduce' to something else. Even though we may know that mental events of the kind which we experience can be produced by the same forces which operate in the rest of nature, we shall never be able to say which are the particular physical events which 'correspond' to a particular mental event.

BIBLIOGRAPHY

In addition to the works quoted in the text this bibliography contains the titles of the works of which I can now remember that they influenced me in the original formulations of the theory here developed (marked with an asterisk), and a few additional works on sensory learning not explicitly referred to in the text. Where the reference is to a translation or edition of a date different from the original, the date of the latter is given first and the former added in brackets.

ADRIAN, E. D., 1947, *The Physical Background of Perception*, Oxford University Press.

ASHBY, W. ROSS, 1945, 'The physical origin of adaptation by trial and error', *J. Gen. Psychol.*, 32, 13–25.

    1946, 'Dynamics of the cerebral cortex. The behavioral properties of the system in equilibrium', *Am. J. Psychol.*, 59, 682– .

    1947a, 'Principles of self-organizing dynamic systems', *J. Gen. Psychol.*, 37, 128ff.

    1947b, 'Dynamics of the cerebral cortex. Automatic development of equilibrium in self-organizing systems', *Psychometrica*, 12.

    1948, 'Design for a brain', *Electronic Engineering*, 20.

    1949, Review of N. Wiener, *Cybernetics*, in *J. Mental Sc.*, 95, 716–724.

ARGELANDER, A., 1927, *Das Farbenhören und der synaesthetische Faktor in der Wahrnehmung*, Jena, Gustav Fischer.

BARGMANN, W., 1947, 'Das Substrat des nervösen Geschehens', *Universitas 2*.

*BECHER, E., 1911, *Gehirn und Seele*, Heidelberg.

*BECHTEREV, W. von, 1913, *Objektive Psychologie*, Leipzig and Berlin.

BECK, LLOYD H. and MILES, WALTER R., 1947, 'Some theoretical and experimental relationships between infrared absorption and olfaction', *Science*, 106, 511.

BERTALANFFY, L. von., 1942, *Theoretische Biologie*, Vol. II. Berlin, Julius Springer.

    1949, *Das biologische Weltbild*, Vol. I, Bern, A. Francke A. G.

    1950a, 'The theory of open systems in physics and biology', *Science*, 23–29.

    1950b, 'An outline of general system theory', *British J. for the Philos. of Sci.*, I: 12, 1–32.

BICHOWSKI, F. R., 1925, 'The mechanism of consciousness: the pre-sensation', *Am. J. Psychol.*, 36, 588–600.

BINNS, H., 1926, 'A comparison between the judgments of individuals skilled in the textile trade and the natural judgments of untrained adults and children', *J. Textile Institute*, 17, 1615–1641.

BIBLIOGRAPHY

BINNS, H., 1937, 'Visual and tactual judgment as illustrated in a practical experiment', *Brit. J. Psychol.*, 27, 404–409.

BLAIR, G. W. SCOTT and COPPEN, F. M. V., 1939, 'The subjective judgment of elastic and plastic properties of soft bodies; the "differential threshold" for viscosities and compression moduli', *Proc. Royal Soc.*, 8, 128, 109–125.

1940, 'The subjective judgment of elastic and plastic properties of soft bodies', *Brit. J. Psychol.*, 31, 61–79.

BORING, EDWIN G., 1933, *The Physical Dimensions of Consciousness*, New York, The Century Co.

1935, 'The relation of the attributes of sensation to the dimensions of the stimulus', *Phil. Science*, 2, 236–245.

1936, 'Psychological systems and isomorphic relations', *Psychol. Rev.*, 43.

1937, 'A psychological function is a relation of successive differentiations of events in the organism', *Psychol. Rev.*, 44, 445–461.

1942, *Sensation and Perception in the History of Experimental Psychology*, New York, Appleton-Century.

BRALY, K. W., 1933, 'The influence of past experience on visual perception', *J. Exp. Psychol.*, 16, 613–643.

BROAD, C. D., 1925, *The Mind and its Place om Nature*, London, Kegan Paul.

BROGDEN, W. J., 1939, 'Sensory pre-conditioning', *J. Exp. Psychol.*, 25, 323–332.

1942, 'Test of sensory pre-conditioning with human subjects', *J. Exp. Psychol.*, 31, 505–517.

1947., 'Sensory pre-conditioning of human subjects', *J. Exp. Psychol.*, 37, 527–539.

1950, 'Sensory pre-conditioning measured by the facilitation of auditory acuity', *J. Exp. Psychol.*

CANNON, W. B., 1932, *The Wisdom of the Body*, London, Kegan Paul.

CAPURSO, A. A., 1934, 'The effect of an associative technique in teaching pitch and interval discrimination', *J. Appl. Psychol.*, 18, 811 f.

CARNAP, R., 1928, *Der logische Aufbau der Welt*, Berlin, Weltkreis Verlag.

1934, 'Logical Foundations of the Unity of Science', *Encyclopaedia of Unified Science*, Vol. I, No. 1, University of Chicago Press.

*CARNERI, B., 1893, *Empfindung und Bewusstsein*, Bonn.

CATTELL, R. B., 1930, 'The subjective character of cognition and the pre-sensational development of perception', *Brit. J. Psychol.*, *Monogr. Suppl.*, 5/14.

CONNETTE, EARLE, 1941, 'The effect of practice with knowledge of results upon pitch discrimination', *J. Educ, Psychol.*, 32, 523–532.

CRAIK, KENNETH J. W., 1943, *The Nature of Explanation*, Cambridge Univ. Press.

CZERMAK, J. N., 1855, 'Weitere Beiträge zur Physiologie des Tastsinnes', *Sitzungsberichte der Kais. Akademie der Wissensch.*, Vienna, XVIII.

DIMMICK, F. L., 1946, 'A color aptitude test, 1940 experimental edition', *J. Appl. Psychol.*, 30, 10–22.

DODGE, RAYMOND, 1931, *Conditions and Consequences of Human Variability*, New Haven, Yale University Press.

DUNCKER, K., 1939, 'The influence of past experience on perceptual properties', *Am. J. Psychol.*, 52, 255–265.

196

BIBLIOGRAPHY

*EDINGER, L., 1909, 'Die Beziehungen der vergleichenden Anatomie zur vergleichenden Psychologie', *Bericht des 3. Kongresses für experimentelle Psychologie*, Leipzig.

EDMONDS, E. M. and SMITH, M. E., 1923, 'The phenomenological description of musical intervals', *Am. J. Psychol.*, 34, 287–291.

ERDMANN, BENNO, 1920, *Grundzüge der Reproduktionspsychologie*, Berlin and Leipzig.

ERISMANN, THEODOR, 1948, 'Das Werden der Wahrnehmung', *Kongress der deutschen praktischen Psychologen*, 1. 61–86.

*EXNER, S., 1894, *Entwurf einer physiologischen Erklärung der psychischen Erscheinungen*, Leipzig and Vienna.

EWERT, P. S., 1930, 'A study of the effects of inverted retinal stimulation upon spatially co-ordinated behavior', *Genet. Psychol. Mong.*, 7, 177–363.

FRIEDLINE, C. L., 1918, 'Discrimination of cutaneous patterns below the two-point limen', *Am. J. Psychol.*, 29, 400–419.

Fundamental Mathematics, 1948, by the College Mathematics Staff of the University of Chicago, 3rd. ed.

GIBSON, J. J., 1941, 'A critical review of the concept of set in contemporary experimental psychology'. *Psychol Bull.*, 38, 781–817.

GILBERT, G. M., 1941, 'Inter-sensory facilitation and inhibition', *J. Gen Psychol.*, 24, 391–407.

GOLDSTEIN, KURT, 1949, *The Organism. A Holistic Approach to Biology Derived from Biological Data in Man*, New York, American Book Company.

GREEN, THOMAS HILL, 1884, *Prolegomena to Ethics*, Oxford University Press.

HARTMANN, G. W., 1935, *Gestalt Psychology, A Survey of Facts and Principles*, New York, Ronald Press.

HAYEK, F. A., 1942–1944, 'Scientism and the Study of Society', *Economica*, N.S., 9, 267–291, 10, 34–63, 11, 27–39, reprinted in *The Counter-Revolution of Science*, (1952). Glencoe, Ill., Free Press.

HAZZARD, F. W., 1930, 'A Descriptive Account of Odors', *J. Exp. Psychol.* 13, 297–331.

HEAD, Henry, 1920, *Studies in Neurology*, London, Hodder & Stoughton.

HEBB, D. O., 1949, *The Organization of Behavior. A Neuropsychological Theory*, New York, Wiley.

*HELMHOLTZ, H. von, 1879, *Die Tatsachen der Wahrnehmung*, Berlin.

1866 (1925), *Helmholtz's Treatise on Physiological Optics*, translated from the third German ed. by James P. C. Southall, Vol. III, *The Perception of Vision*. The Optical Society of America.

HENNING, HANS, 1917, 'Versuche über Residuen', *Z. f. Psychol.*, 78, 189–269.

1919, 'Die assoziative Mitwirkung, das Vorstellen von noch nie wahrgenommenen und deren Grenzen', *Z. f. Psychol.*, 81, 1–96.

1922, 'Assoziationsgesetz und Geruchsgedächtnis', *Z. F. Psychol.*, 89, 38–80.

1924, *Der Geruch, Ein Handbuch*, Leipzig, J. A. Barth.

1927, 'Psychologische Studien am Geschmackssinn' and 'Psychologische Studien am Geruchsinn', *Handbuch der biologischen Arbeitsmethoden*, ed. E. Abderhalden, Abt. VI, Teil A, 627–740 and 741–836.

BIBLIOGRAPHY

HENRI, Victor, 1898, *Die Raumwahrnehmungen des Tastsinnes*, Berlin, Reuther und Reichhard.

*HERING, EWALD, 1870, *Ueber das Gedächtnis als eine allgemeine Funktion der organisierten Materie*, Vienna.

1885 (1913), 'Ueber die spezifischen Energien des Nervensystems', *Lotos*, IV. F. 5f (Prague), English translation in *Memory. Lectures on the specific energies of the nervous system*, 4th ed. Chicago, Open Court Publishing Co.

HERRICK, C. JUDSON, 1926, *Brains of Rats and Men*.

HILGARD, E. R., CAMPBELL, R. K. and SEARS, W. N., 1937, 'Conditioned Discrimination with and without verbal report', *Am. J. Psychol.*, 49, 564–580.

HILGARD, E. R. and MARQUIS, D. G., 1940, *Conditioning and Learning*, New York, Appleton-Century.

HILLARP, N. A., 1947, Structure of the synapse', *Acta Anatomica* 2, Suppl.

HOLT, EDWIN B., 1937, 'Materialism and the criterion of the psychic', *Psychol. Rev.*, 44, 33–57.

HORNBOSTEL, E. M. von, 1926, 'Unity of the Senses', Psyche 7, 83–89.

1931, 'Ueber Geruchshelligkeit', *Arch. f. ges. Physiol.* 227.

HULL, C. L., 1943, *Principles of Behavior*, New York, Appleton-Century.

HUMES, J. F., 1930, 'The effect of practice upon the upper limen of tonal discrimination', *Am. J. Psychol.* 42.

HUMPHREY, George, 1932, *The Story of Man's Mind*, rev. and enl. ed., New York, Dodd, Mead & Co.

*JAMES, WILLIAM, 1890, *Principles of Psychology*, London, Macmillan.

JENNINGS, H. S., 1906, *The Behaviour of Lower Animals*, New York, Columbia University Press.

*JODL, FRIEDRICH, 1916, *Lehrbuch der Psychologie*, 4th ed., Stuttgart.

KENNETH, J. H., 1927, 'An experimental study of affects and associations due to odors, *Psychological Monogr.* 37/2.

KINGSLEY, H. L., 1946, *The Nature and Conditions of Learning*, New York, Prentice Hall.

KLEINT, H., 1937–1940, 'Versuche über Wahrnehmung', *Z. f. Psychol.*, 140, 109–138; 142, 259–290; 142, 299–317; 148, 145–204; 149, 31–82.

KLUEVER, H., 1931, 'The equivalence of stimuli in the behavior of monkeys', *J. Gen. Psychol.*, 39.

1933, *Behavior Mechanism of Monkeys*, Univ. of Chicago Press.

1935, 'The study of personality and the method of equivalent and non-equivalent stimuli', *Character and Personality*, 5.

1949, 'Psychology at the beginning of World War II: Meditations on the impending dismemberment of Psychology', *J. Psychol.* 28.

KOEHLER, W., 1913, 'Ueber unbemerkte Empfindungen und Urteilstäuschungen', *Z. f. Psychol.*, 66, 51–80.

1920, *Die physischen Gestalten in Ruhe und im stationärem Zustand*.

1929, *Gestalt Psychology*.

KOEHLER, W. and HELD, R., 1949, 'The cortical correlate of pattern vision', *Science*.

KOFFKA, K., 1929, 'On the structure of the unconscious', in *The Unconscious. A Symposion*, New York, Knopf.

1935, *Principles of Gestalt Psychology*, London, Kegan Paul.

BIBLIOGRAPHY

*KRIES, J. von, 1898, *Ueber die materiellen Grundlagen der Bewusstseinserscheinungen*, Freiburg i. B.

1923, *Allgemeine Sinnesphysiologie*, Leipzig, F. C. W. Vogel.

*KRUEGER, FELIX, 1915, *Ueber Entwicklungspsychologie, ihre sachliche und geschichtliche Notwendigkeit*, Leipzig, W. Engelmann.

KUELPE, O., 1893 (1895), *Outlines of Psychology*, London-New York.

LASHLEY, K. S., 1923, 'The behavioristic interpretation of consciousness', *Psychol. Rev.*, 30.

1929, *Brain Mechanism and Intelligence*, Chicago, Univ. of Chicago Press.

1934, 'Nervous mechanism in learning', *Handbook of General Experimental Psychology*, ed. C. Murchison.

1942, 'The problem of cerebral organization in vision', *Visual Mechanisms*, ed. H. Klüver, *Biol. Sympos.*, 7.

LEEPER, R. W., 1935, 'A study of a neglected portion of the field of learning—the development of sensory organization', *Pedagogical Seminary and J. Gen. Psychol.*, 46, 41–75.

LEWES, G. H., 1874–1879, *Problems of Life and Mind*, London.

LOCKE, John, 1690, *An Essay Concerning Human Understanding*.

LORENZ, K., 1943, 'Die angeborenen Formen möglicher Erfahrung', *Z. f. Tierpsychol.*, 5, 235–409.

McCLEARY, R. and LAZARUS, R. S., 1949, 'Autonomic discrimination without awareness: an interim report', *J. of Personality*, 18, 170–179.

McCULLOCH, W. S., 1948, 'A recapitulation of the theory with a forecast of several extensions', *Teleological Mechanisms*, ed. L. K. Frank, *Annals of the New York Academy of Science*, 50, 259ff. See also PITTS, W.

McDOUGALL, W., 1923, *Outline of Psychology*, London, Methuen.

McGEOCH, JOHN A., 1936, 'The vertical dimensions of mind', *Psychol. Rev.* 43, 107–129.

*MACH, ERNST, 1885, *Die Analyse der Empfindungen*, Jena, G. Fischer.

1905, *Erkenntnis und Irrtum*, Leipzig, J. A. Barth.

*MAHLING, F., 1926, 'Das Problem der "Audition colorée" ', *Arch. f. d. ges. Psychol.* 57, 165—302.

MARGENAU, HENRY, 1950, *The Nature of Physical Reality*, New York, McGraw-Hill.

MARTIN, L. J. and MUELLER, G. E., 1899, *Zur Analyse der Unterschiedsempfindlichkeit: experimentelle Beiträge*, Leipzig, J. A. Barth.

MEES, C. E. KENNETH, 1946, *The Path of Science*, New York, Wiley.

METZGER, WOLFGANG, 1941, *Psychologie: Die Entwicklung ihrer Grundannahmen seit der Einführung des Experiments*. (Wissenschaftliche Forschungsberichte. Naturwissenschaftliche Reihe, Band 52) Dresden und Leipzig, Theodor Steinkopf.

MILL, JAMES, 1829 (1869), *Analysis of the Phenomena of the Human Mind*, a new edition, ed. by J. S. Mill, London.

MILLER, J. G., 1939, 'Discrimination without awareness', *Am. J. Psychol.* 52, 562–578.

1940, 'The role of motivation in learning without awareness', *Am. J. Psychol.*, 53, 229–239.

1942, *Unconsciousness*, New York, Wiley.

BIBLIOGRAPHY

MINER, J. B., 1905, 'A case of vision acquired in adult life', *Psychol. Rev.Monogr.* 6/6, 103–118.

MONCRIEFF, R. W., 1946, *The Chemical Sense*, New York, John Wiley.

MOORE, H. T., 1914, 'The genetic aspects of consonance and dissonance', *Psychol. Monogr.*, 17/73.

MORGAN, C. T., 1943, *Physiological Psychology*, New York, McGraw Hill.

MÜLLER, JOHANNES, 1838, *Handbuch der Physiologie des Menschen.*

NEURATH, O., 1933, *Einheitswissenschaft und Psychologie*, Vienna, Gerold.

NORTHWAY, MARY L., 1940, 'The concept of the "schema" ', *Brit. J. Psychol.* 31, 22 f.

NOYES, C. R., 1950, 'What kind of psychology does economics need?', *Canadian J. of Economics and Pol. Sci.*, 16, 210–215.

OGDEN, R. M., 1926, *Psychology and Education.*

PEAK, H., 1933, 'An evaluation of the concepts of reflex and voluntary action', *Psych. Rev.*, 40, 71–80.

PETERSON, Joseph, 1933, 'Aspects of learning', *Psychol. Rev.*, 42, 1–27.

PETROVITCH, M., 1921, *Mécanisms communs aux phénomènes disparates*, Paris, Félix Alcan.

PIAGET, JEAN, 1942, *Classes, Relations et Nombres. Essai sur les 'Groupements' de la logistique et la réversibilité de la pensée*, Paris, Vrin.

1947 (1950), *The Psychology of Intelligence*, London, Routledge & Kegan Paul.

1949, *Traité de Logique*, Paris, Armand Colin.

PITTS, W. and McCULLOCH, 1947, 'How we know universals. The perception of auditory and visual forms', *Bull. Math. Biophysics*, 9, 127–147.

PLANK, M., 1926, *A Survey of Physics*, London.

1942 (1949), 'The meanings and limits of exact science', in *Scientific Autobiography*, New York, 80–120.

PRATT, CARROL C., 1939, *The Logic of Modern Psychology*, New York.

RENSHAW, S., 1930, 'The errors of cutaneous localization and the effect of practice on the localizing movement in children and adults'. *J. Genet. Psychol.*, 38, 223–238.

RENSHAW, S., WHEERY, R. J. and NEWLIN, J. C., 1930, 'Cutaneous localization in congenitally blind versus seeing children and adults', *J. Genet. Psychol.*, 38, 239–294.

REVESZ, G., 1924, 'Experiments on animal space perception', *7th Int. Congr. of Psychol. Proceedings and Papers*, 29–56.

RIES, G., 'Untersuchungen über die Sicherheit der Aussage', *Z. f. Psychol.*, 88, 145–204.

RIESEN, A. H., 1942, 'The development of visual perception in man and chimpanzee', *Science*, 106, 107–108.

1950, 'Arrested vision', *Scientific American*, 183.

RIKER, B. L., 1946, 'The ability to judge pitch', *J. Exp. Psychol.*, 36, 331–346.

ROBINSON, E. S., 1931, *Association Theory Today*, New York, Century.

ROSENBLUETH, A., WIENER, N. and BIGELOW, J. H., 1943, 'Behavior, purpose, and teleology', *Philos of Science.*, 10, 18–23.

RUSSELL, BERTRAND, 1921, *Analysis of Mind*, London, Allen & Unwin.

1927, *The Analysis of Matter*, London, Allen & Unwin.

BIBLIOGRAPHY

RYAN, T. A., 1940, 'Interrelations of sensory systems in perception', *Psychol. Bull.*, 37, 659–698.

RYLE, G., 1942, 'Knowing how and knowing that', *Proceed. Arist. Soc.*, N.S., 46.
1949, *The Concept of Mind* London, Hutchinson.

SCHILLER, M., 1932, 'Die Rauhigkeit als intermodale Erscheinung', *Z. f. Psychol.*, 127, 265–298.

*SCHLICK, M., 1918, *Allgemeine Erkenntnislehre*, Berlin, J. Springer.

SCHNABEL, 1881, 'Beiträge zu der Lehre von der Schlechtsichtigkeit durch Nichtgebrauch der Augen', *Berichte des naturw.-mediz. Vereins in Innsbruck*, 11, 32–59.

SCHUMANN, F., 1908, *Beiträge zur Analyse der Gesichtswahrnehmungen*, Leipzig.
1922, 'Das Erkennungsurteil', *Z. f. Psychol.*, 88, 205–224.

*SEMON, R. W., 1904, *Die Mneme als erhaltendes Princip im Wechsel des organischen Geschehens*, Leipzig.
1909, *Die Mnemischen Empfindungen*, Leipzig.

*SENDEN, M. von, 1932, *Raum- und Gestaltauffassung bei operierten Blindgeborenen vor und nach der Operation*, Leipzig, J. A. Barth.

SHERRINGTON, Sir CHARLES, 1933, *The Brain and its Mechanism*, Cambridge University Press.
1940, *Mann: On His Nature*, Cambridge University Press.
1949, 'Mystery of mysteries; The human brain', *New York Times Magazine*, Dec. 4.

SKRAMLIK, EMIL von, 1937, *Psychophysiologie des Tastsinnes* (Suppl. 4 to *Arch. f.d. ges. Psychol.*)

SPENCE, K. W., 1944, 'The nature of theory construction in contemporary psychology', *Psychol. Rev.*, 51, 47–68.
1948, 'The postulates and methods of "Behaviorism"', *Psychol. Rev.*, 55, 67–78.

STERN, WILLIAM, 1938, *General Psychology from the Personalistic Standpoint*, New York.

STEVENS, S. S., 1934, 'The attributes of tones', *Proc. Nat. Acad. Sci.*, 20, 457–459.
1935a, 'The operational definitions of psychological concepts', *Psychol. Rev.*, 42, 517–527.
1935b, 'The operational basis of psychology', *Am. J. Psychol.*, 47, 323–330.
1936, 'Psychology the propedeutic science', *Philos. Science*, 3, 90–103.
1939a, 'Psychology and the science of science', *Psycho. Bul.* 36, 221–263.
1939b, 'On the problem of scales for the measurement of psychological magnitudes', *J. Unif. Sci.*, 9, 94–99.
1946a, 'On the theory of scales of measurement', *Science*, 103, 677–680.
1946b, 'The two basic mechanisms of sensory discrimination', *Fed. Am. Soc. Exp. Biol., Proc.*, 5, part 2, 101.
1948, 'Sensation and psychological measurement' in E. G. Boring, H. S. Langfeld and H. P. Weld (eds.), *Foundations of Psychology*, New York, Wiley.

STEVENS, S. S. and DAVIS, H., 1938, *Hearing: Its Psychology and Physiology*, New York, Wiley.

STEVENS, S. S., MORGAN, C. T. and VOLKMAN, J., 1941, 'Theory of the neural quantum in the discrimination of pitch', *Am. J. Psychol.*, 54, 315–353.

BIBLIOGRAPHY

STEVENS, S. S., and VOLKMANN, J., 1940a, 'The quantum theory of sensory discrimination', *Science*, 92, 583–585.

1940b, 'The relation of pitch to frequency: a revised scale', *Am. J. Psychol.*, 54, 329–353.

*STÖHR, A., 1917, *Psychologie*, Vienna, Braumuller.

STOUT, G. F., 1915, *Manual of Psychology*, 3rd. ed., London, University Tutorial Press.

STRATTON, G. M., 1897, 'Vision without inversion of the retinal image', *Psychol. Rev.* 4, 341–360, 463–481.

THOMPSON, D. W., 1942, *On Growth and Form*, a new ed., Cambridge University Press.

THORNDIKE, EDWARD L., 1913, *The Psychology of Learning (Educational Psychology*, vol. III) New York, Teachers College, Columbia University Press.

TITCHENER, E. B., 1905, *Experimental Psychology*, II/I.

TOLMAN, E. C., 1932, *Purposive Behavior in Animals and Men*, New York, Century.

1948, 'Cognitive maps in rats and men', *Psychol. Rev.*, 55, 189–208.

1949, 'There is more than one kind of learning', *Psychol. Rev.*, 56, 144–155.

TROLAND, L. T., 1928, *The Fundamentals of Human Motivation*, New York, D. Van Nostrand & Co.

1930, *The Principles of Psychophysiology*, Vol. II, New York, D. Van Nostrand & Co.

UNDERWOOD, B. J., 1949, *Experimental Psychology*, New York, Appleton-Century.

*URBANTSCHITSCH, V. 1888, 'Ueber die Wechselwirkung zwischen verschiedenen Sinnesempfindungen', *Pflüger's Archiv.*, 42.

*VERWORN, M., 1907, *Die Mechanik des Geisteslebens*, Leipzig, Teubner.

1920, *Die Entwicklung des menschlichen Geistes*, 4th ed., Jena, G. Fischer.

VOLKMANN, A., 1858, 'Ueber den Einfluss der Uebung auf das Erkennen, räumlicher Distanzen', *Berichte über die Verh. d. kgl. Sächs. Gesellsch. der Wissensch. zu Leipzig*, 10, 38–76.

WALLS, G. L., 1942, *The Vertebrate Eye and Its Adaptive Radiation*, Bloomfield Hills, Mich., The Cranbrook Institute of Science.

WALTER, W. GRAY, 1950, 'An Imitation of Life', *Scientific American*, 182/5.

WEDELL, C. H., 1942, 'The nature of the absolute judgment of pitch', *J. Exp. Psychol.*, 30, 426–431.

WEISS, A. P., 1925, *A Theoretical Basis of Human Behavior*, Columbus, Ohio, R. G. Adams & Co.

WEISS, PAUL, 1941, 'Self-differentiation as the Basic Pattern of Co-ordination', *Comparative Psychology Monographs*, XVII/4 (88).

WEIZSAECKER, V. von, 1940, *Der Gestaltkreis. Theorie der Einheit von Wahrnehmen und Bewegen*, Stuttgart, Georg Thieme.

WELLEK, A., 1931, 'Zur Geschichte und Kritik der Synaesthesie-Forschung', *Arch. f. ges. Psychol.*, 79.

WERNER, H., 1926a, 'Mikromelodik und Mikroharmonik', *Zeitsch. f. Psychol.*, 98, 74–89.

1926b, 'Die Ausprägung von Tongestalten', *Zeitsch. f. Psychol.*, 101, 159–181.

1930, 'Untersuchungen über Empfindung und Empfinden', *Z. f. Psychol.* 114, 152–166.

BIBLIOGRAPHY

WERNER, H., 1948., *Comparative Psychology of Mental Development*, Chicago, Follet Publ.

WIENER, N., 1948a, *Cybernetics*, New York, Wiley.

    1948b, 'Time, Communications, and the Nervous System', in *Teleological Mechanisms*, ed. L. K. Frank, *Annals of the New York Academy of Sciences*, 50, 197–278.

    See also ROSENBLUETH, A.

WILLIAMS, J. T., 1922, 'Extraordinary development of the tactile and olfactory senses compensatory for the loss of sight and hearing', *J. Am. Med. Ass.*, 79, 1331–1334.

WINSLOW, C. N., 1933, 'Visual illusions in the chick', *Archiv. of Psychol.*

WOODGER, J. H., 1929, *Biological Principles*, London, Kegan Paul.

WOODWORTH, R. S., 1938, *Experimental Psychology*, New York, Henry Holt.

*WUNDT, W., 1902–3, *Grundzüge der physiologischen Psychologie*, 5th ed., Leipzig, Engelmann.

WYATT, R. F., 1945, 'Improvability of pitch discrimination', *Psychol. Monogr.*, 58/2 (No. 267).

YOUNG, P. T., 1928, 'Auditory localization with acoustic transposition of ears', *J. Exp. Psychol.*, 11, 399–429.

NOTE

While there would be little point in adding to this bibliography the titles of the various relevant American and English works which have appeared since the middle of 1950, when the final text of the present book was completed, I should like to draw special attention to the full account of some interesting experiments on sensory learning (of which I then knew only through the preliminary accounts in T. Erisman, 1948) given in:

KOHLER, Ivo, 1951, *Ueber Aufbau und Wandlungen der Wahrnehmungswelt. Insbesondere über 'bedingte Empfindungen'*. Oesterreichische Akademie der Wissenschaften, Philosophisch-Historische Klasse, *Sitzungsberichte*, 227/1, 1–118.

INDEX